Mathematics Curriculum in Pacific Rim Countries— China, Japan, Korea, and Singapore

Proceedings of a Conference

a volume in
Research in Mathematics Education

Series Editor:
Barbara J. Dougherty
University of Mississippi

Research in Mathematics Education

Barbara J. Dougherty, Series Editor

*Mathematics Curriculum in Pacific Rim Countries—China, Japan, Korea, and
Singapore: Proceedings of a Conference* (2008)
edited by Zalman Usiskin and Edwin Willmore

The History of the Geometry Curriculum in the United States (2008)
by Nathalie Sinclair

The Classification of Quadrilaterals: A Study in Definition (2008)
edited by Zalman Usiskin and Jennifer Griffin

*The Intended Mathematics Curriculum as Represented in
State-Level Curriculum Standards: Consensus or Confusion?* (2006)
edited by Barbara Reys

Mathematics Curriculum in Pacific Rim Countries— China, Japan, Korea, and Singapore

Proceedings of a Conference

edited by

Zalman Usiskin and Edwin Willmore
The University of Chicago

Information Age Publishing, Inc.
Charlotte, North Carolina • www.infoagepub.com

Library of Congress Cataloging-in-Publication Data

Mathematics curriculum in Pacific rim countries--China, Japan, Korea, and Singapore: proceedings of a conference / edited by Zalman Usiskin and Edwin Willmore.

   p. cm. — (Research in mathematics education)

  Includes bibliographical references.

   ISBN 978-1-59311-953-9 (pbk.) — ISBN 978-1-59311-954-6 (hardcover)

1. Mathematics—Study and teaching—East Asia—Congresses. 2. Mathematics—Study and teaching—Singapore—Congresses. 3. Mathematics—Study and teaching—Cross-cultural studies—Congresses. 4. Curriculum evaluation—Congresses. I. Usiskin, Zalman. II. Willmore, Edwin.

QA14.E18M38 2008
510.71'05—dc22

2008027249

Organization of the CSMC First International Mathematics Curriculum Conference and the publication of these proceedings were made possible by the generous support of the National Science Foundation, Grant No. ESI-0333879 to the Center for the Study of Mathematics Curriculum, Barbara Reys, director. Opinions, findings and conclusions presented here are those of the authors and do not necessarily reflect the opinions of others at the center or of the National Science Foundation.

Printed in the United States of America

CONTENTS

PART III:
U.S. PERSPECTIVES ON THE
CURRICULA OF ASIAN PACIFIC RIM COUNTRIES

PART IV:
STATUS OF CALCULATOR AND
COMPUTER TECHNOLOGY IN THE K–12 CURRICULUM

PART V:
ROLE OF TESTING IN THE K–12 CURRICULUM

PART VI:
REFLECTIONS BY CONFERENCE ATTENDEES

PREFACE

Zalman Usiskin

This volume contains the proceedings of the First International Curriculum Conference sponsored by the Center for the Study of Mathematics Curriculum (CSMC), held November 8–10, 2005 on the campus of the University of Chicago.

The CSMC is one of the National Science Foundation Centers for Learning and Teaching (Award No. ESI-0333879). As noted on the CSMC Web site (http://mathcurriculumcenter.org), the CSMC serves the K–12 educational community by focusing scholarly inquiry and professional development around issues of mathematics curriculum. Major areas of work include understanding the influence and potential of mathematics curriculum materials, enabling teacher learning through curriculum material investigation and implementation, and building capacity for developing, implementing, and studying the impact of mathematics curriculum materials. The work of the CSMC is not driven by a particular philosophy or ideology. As researchers, CSMC staff try to maintain a healthy skepticism throughout all phases of the center's activities considering both international and multiple U.S. perspectives with an ultimate goal to produce research-based knowledge and products that will enlighten and serve the range of users of mathematics curriculum materials.

Mathematics Curriculum in Pacific Rim Coutries—China, Japan, Korea, and Singapore: Proceedings of a Conference, pp. ix–xi
Copyright © 2008 by Information Age Publishing

The venue, the University of Chicago, had been the location of international conferences on mathematics education in 1985, 1988, 1991, and 1999. This conference differed from those in that it was focused on a small number of countries from the same general region of the globe and on a single topic: the ways in which those countries create their mathematics education curricula. The countries—China, Japan, Korea, and Singapore (in alphabetical order, which also happens to be the order of their populations)—have each been in the news because of their performance on international tests and/or their economic performance and potential. They also have centralized education ministries that create a single mathematics curriculum framework followed in the entire country.

We decided to invite two speakers from each of these countries, one who was in on the creation of their national mathematics curriculum framework and another who had been in the position of writing textbooks to follow that framework, eight speakers in all. We asked these speakers to tell us about the processes of creation and the issues that they needed to consider in their work.

Then, following these presentations, we invited two researchers from the United States whose work involved comparing U.S. curricula with the curriculum of one of these countries, again giving us eight speakers in all.

All of the speakers from outside the United States were on one of two conference panels, discussing testing or technology. Also on these panels were two or three others from the United States who conduct research in these areas.

In this volume are papers from all of the speakers at the conference.

Following the conference, doctoral students from the CSMC home universities and several other institutions gathered for a 1.5 days' postconference to discuss what they had heard. We invited the students to write papers, either singly or in groups, and indicated that we would pick the best of these papers for publication. Two groups of students wrote papers and we felt that both were worthy, so both are included in this volume.

As is almost always the case at conferences like these, there are surprises. Not wanting to preempt anyone's reading of the conference papers, I will content myself with one thing from each country that surprised me and one, perhaps surprising, commonality among the countries. Large numbers of students in China finish their formal education at Grade 9. One suspects that this may provide pressure to cover more mathematics before Grade 8 than might otherwise be the case. Parents in Korea spend more money on afterschool tutoring classes for their children than the government spends on education. The amount of time in these classes may itself account for the high mathematics scores on international comparison tests. In both Japan and Korea, there is great concern about the negative view that students have toward mathematics compared to students in other countries that may not perform as well. At the end of fourth grade in Singapore, students are held back if they cannot keep up with their classmates. In all these countries, curricula are differentiated for students with different interests, usually around Grade 10 or 11.

We think the reader will agree that the papers are of very high quality.

ACKNOWLEDGMENTS

A conference of this type requires the work of many people and the support of many organizations. We thank first the National Science Foundation, without whose support this conference could not have taken place.

We thank the many speakers from the United States and abroad, who shared their thoughts and expertise with us, and the conference participants who contributed to the conversation and ambience of the meeting.

The planning of the conference was done by a CSMC committee consisting of: Glenda Lappan, Ira Papick, Robert Reys, Zalman Usiskin, and Steven Ziebarth, overseen by the CSMC Leadership Team. The local arrangements for the conference were led by Carol Siegel and carried out by staff of the University of Chicago School Mathematics Project.

This volume was edited by Edwin Willmore and Zalman Usiskin. We wish also to thank Barbara Dougherty for her role in bringing these proceedings to publication, and to Information Age Publishing for publishing the manuscript.

CHAPTER 1

INTRODUCTIONS

The opening session of the conference took place in the Max Palevsky auditorium of Ida Noyes Hall. The nearly 200 attendees were welcomed by the chair of the conference planning committee, the dean of the Division of Social Sciences at the University of Chicago, and one of the principal investigators of the Center for the Study of Mathematics Curriculum. The following is a transcription from the opening session.

Zalman Usiskin

As the chair of the planning committee and the local host, it is my honor and pleasure to welcome you the to the First International Curriculum Conference organized by the Center for the Study of Mathematics Curriculum. We're delighted to have you here on such a beautiful day.

Curriculum researchers generally in education, and within mathematics education specifically, have come to distinguish among several types of curricula. There is the ideal curriculum, as described in guidelines and frameworks of what we would like teachers to teach and students to learn. The implemented curriculum is what teachers teach and students encounter in school. The textbook curriculum consists of the materials that teachers have available to use. The tested curriculum is the curriculum as judged from what appears on tests given to students. And the achieved curriculum is what students learn.

Mathematics Curriculum in Pacific Rim Coutries—China, Japan, Korea, and Singapore:
Proceedings of a Conference, pp. 1–5

Although we often think of these curricula as nested within each other, the implemented and textbook curricula being a part of the ideal and the achieved being part of the tested, the relationships among these curricula are quite a bit more complex. Studying these curricula requires quite different research programs. Ideal curricula, the textbook curriculum, and the tested curriculum can often be analyzed in the ivory tower without ever talking to a child or visiting a school. On the other hand, to study the implemented curriculum requires classroom observation, and often, interviews of teacher and student. And to examine the achieved curriculum requires all of the complexities that testing and the analysis of test results entails, and often means looking beyond schools to the workplace and to society at large.

All of these curricula have traditions. We know that even for a subject like mathematics, often considered to be universal, these traditions differ from country to country. It is probably safe to say we are all here because we believe that learning what is done in other countries brings perspective on what we do in our country, provides ideas and cautions for possible directions we might take, and gives us opportunities to think together with people about the same issues that we study, with the ultimate purpose for all of us of making this world a better place for its citizens to live.

The focus of this conference is on the processes that lead to two of the types of curricula that I have mentioned: the ideal curriculum and the textbook curriculum. We have invited speakers, from four countries, who have been intimately involved in the creation of guidelines or textbooks, in some cases both, to discuss with us the assumptions that are made and the processes that their country goes through in establishing standards or national curriculum and the ways in which these standards are translated into text materials by authors and implemented in schools. They are joined by speakers from the United States who have compared U.S. practice with those of the participating Asian countries. Also, we have organized panels on technology and testing to talk about these important curricular issues.

We are pleased that John Mark Hansen, dean of the Division of Social Sciences at the University, has been able to take some time out of his very busy schedule to welcome us. John Mark Hansen is the Charles L. Hutchinson Distinguished Service Professor in Political Science. He is one of the nation's leading scholars of American politics, particularly public opinion, interest groups, and elections. Public opinion interest groups have some impact on mathematics curricula so his interests aren't as far from mathematics education as some might think. Mark joined the university faculty in 1986 as an assistant professor in political science and was quickly promoted. He served as chair of political science from 1995 to 1998 and was associate provost for education and research from 1998 to 2001. In 1999 he was named the William R. Keening, Jr. Professor in Political Science in the college. During the 2001-2002 academic year he left us to be professor of government at Harvard University but he came back in 2002 to begin a 5-year term as dean of the Division of Social Sciences at the university. He teaches courses on American government and politics, including classes on the United States Congress, interest groups, and elections. He also participates regularly in the democracy and social science core sequence in the college: Dean Hansen.

John Mark Hansen

Thank you. Welcome to Chicago. Welcome to the University of Chicago campus. It's wonderful to have you here. I know that this is an unusually distinguished group and that this is an unusually international group. We're very happy to have you on our campus, very happy that you'll be taking part in the activities over the next 3 days or so.

As I'm sure many of you know the University of Chicago has a very distinguished history in the study of mathematics pedagogy and mathematics education. J. W. A. Young held the title in the very early part of the university's history, the title of professor of pedagogy of mathematics education. It was either the first or the second such position that was created at any American university. In 1902, E. H. Moore, who was then the chair of our mathematics department, gave a famous presidential address to the American Mathematical Society that appealed for improvement in mathematics education and had an electrifying effect on the development of new approaches to mathematics pedagogy in the United States. Through the twentieth century, with figures like Ernst Breslich and Morris Hartung, the university continued its tradition of excellence in the nationwide leadership in mathematics education, and more recently in the late 1940s and early 1950s mathematics professors in the college (including one who is here today, Izaak Wirszup) continued this proud tradition and developed the modern program that was influential in the new mathematics program, SMSG.

I'm sure all of you know about the most recent history which is quite diverse in our activities in math education, ranging from the activities of the math department of Paul Sally and Bob Fefferman in training Chicago area schoolteachers to be more effective at mathematics pedagogy and including the work of the University of Chicago School Mathematics Project in creating curriculum and teacher education in all of the grades, kindergarten through 12. As I'm sure all of you know, all of these activities have been enormously influential in the practice of mathematics pedagogy in the United States and I can give two personal illustrations of that.

One is that my cousin, who is a teacher in Kansas, inquired one time whether I know anything about this UCSMP. At kind of the opposite extreme, the university has been deeply involved in a charter school in the North Kenwood/Oakland neighborhood directly to the north of Hyde Park and Kenwood, and that charter school, the North Kenwood/Oakland School, has adopted the University of Chicago School Mathematics Program for its mathematics pedagogy and has been enormously successful with it.

I see by the program that all of you have a very interesting schedule ahead of you and certainly a very ambitious schedule ahead of you. We're very proud to have you here. We hope you will have some time to see our campus and to enjoy some of the area while you are here. I sense the kind of energy that I know that this group will have and is very true to the energy of the University of Chicago. Thank you very much for being here and welcome.

Zalman Usiskin

Thank you, Mark, for such appropriate remarks. Now I have the pleasure of introducing Barbara Reys. Barbara has the title of distinguished professor of mathematics education at the University of Missouri. She has served on the board of directors of the National Council of Teachers of Mathematics and as a writing group leader for the NCTM's *Principles and Standards of School Mathematics* published in 2000. She has taught elementary, middle, and senior high school and currently works to prepare future middle school mathematics teachers. Her current research focuses on the role and influence of official curriculum documents, including state-level curriculum frameworks and district-adopted mathematics textbooks. Barbara is officially one of the co-PIs of the CSMC. But all of us associated with the Center for the Study of Mathematics Curriculum know her to be much more. She is our leader; our organizer; our prodder—she makes sure that everything gets done: Barbara.

Barbara Reys

Thank you, Zalman. I'd like to welcome you to the first International Conference on Mathematics Curriculum organized by the Center for the Study of Mathematics Curriculum. The center serves the educational community by focusing scholarly inquiry and leadership development around issues of mathematics curriculum. We do this through a coordinated set of activities, including research on the nature and impact of curriculum, doctoral programs with emphasis on curriculum and school-based curriculum leadership development. The center includes faculty and doctoral students at Michigan State University, the University of Missouri-Columbia, and Western Michigan University, staff at the University of Chicago and Horizon Research, and K–12 school partners in Missouri and Michigan. We share the belief that curriculum matters and that by examining the role and influence of curriculum on both teaching and learning and using this information to develop research-based curriculum materials and professional development strategies we can improve student-learning opportunities. We are also committed to preparing the next generation of mathematics curriculum developers, professional development leaders who use curriculum as a central focus of teacher study, and curriculum researchers who generate knowledge to produce the strong curriculum materials. I invite you to learn more about the CSMC by visiting our Web site.

At this conference we focus on the design and development of K–12 mathematics curricula in four countries: China, Japan, Korea, and Singapore. We consider the ways that these countries determine their courses of study and the translation of curriculum standards into materials for teachers and students. Our goal is to stimulate discussion among participants regarding ways to support the development of curriculum frameworks and materials to improve student-learning opportunities in U.S. schools. There are about 200 conference participants, representing 28 different U.S. states, and at least six different countries. Conference participants include university faculty, researchers, K–12 teachers and administrators, Department of Education staff, curriculum developers, and publishers. We think it's the right blend of people and expertise to provide a stimulating and productive conversation.

I want to extend my sincere thanks to all of the speakers who have traveled halfway around the world to join us in this discussion, and also to the staff and faculty here at the University of Chicago, in particular Zalman Usiskin and his staff, including Carol Siegel, who have worked tirelessly to take care of all of the details of the conference so that we can sit back, enjoy, learn, and participate.

PART I

EDUCATION MINISTRY PERSPECTIVES ON K–12 MATHEMATICS CURRICULUM

CHAPTER 2

EDUCATION MINISTRY PERSPECTIVES ON MATHEMATICS CURRICULUM IN JAPAN

Shigeo Yoshikawa
Ministry of Education, Culture, Sports, Science and Technology, Japan

RECENT SITUATIONS

Japan's latest version of the Curriculum Standards was revised by the Ministry of Education, Culture, Sports, Science, and Technology in 1998 (MEXT, 1998a, 1998b, 1998c, 1999a, 1999b, 1999c, 1999d). This latest version of the Curriculum Standards has been implemented in schools since April 2002. The Curriculum Standards are part of legislation by the Ministry of Education. All schools from elementary through secondary levels make their own school curriculum according to the Curriculum Standards. When it comes to mathematics content, all teaching content described in the Standards must be taught to all students. In this sense the standards have a characteristic of "minimal standards."

Mathematics Curriculum in Pacific Rim Coutries—China, Japan, Korea, and Singapore:
Proceedings of a Conference, pp. 9–22

We can see rather a big change in the mathematics sections of the latest standards compared to the previous version. This change comes partly from the reduction of the total number of lesson hours as well as the reduction of teaching contents described in the standards. The reduction of total number of lesson hours was inevitable because we were planning to introduce a complete 5-day school week system at the same time as implementation of the new Curriculum Standards in 2002. On top of that, one of the basic philosophies of the curriculum reform was that students should learn less content, but more expertly in the classroom.

Speaking of the change of teaching contents at the elementary school level, some content, including the symbols of inequality, paper and pencil computation with large integers or complicated decimals, and congruence of geometrical figures, were deleted or moved to upper grade levels. At lower secondary school level, for example, the formula for solving quadratic equations was moved to the upper secondary school mathematics. One of the basic ideas of the revision of the mathematics sections in the Curriculum Standards was that we should carefully select the most important basic mathematics content which students need to live effectively in society and to continue their mathematics learning towards higher levels. Even though students learn less mathematics content, they should be able to understand mathematical meanings more deeply in the process of doing a lot of activities related to mathematics, and should become more willing to learn mathematics.

Right after the latest version of the Curriculum Standards was released in 1998, it seemed that a majority of people in society were in favor of that change. However, opposite opinions gradually appeared in newspapers and other mass media. Some parents started to express anxiety about the school education given to their children.

In January 2002 the Ministry of Education created a booklet titled *Encouragement of Learning* (MEXT, 2002a) for people in the society at large, including schoolteachers. This booklet informs readers that school is a place where students get good academic prowess and develop their rich and healthy minds. The booklet also explains several concrete measures to teach students and to help them achieve such academic prowess. Examples of these concrete measures include:

- Take good care of each individual student and understand them to help them get basic skills and thinking ability. Make use of class according to ability (track class, streaming).

- Prepare opportunities for learning in and out of school. Students can find a space and time to read books in school in the morning before the class. If a student had a hard time during the class, he or she should get appropriate material for complementary learning under the guidance of school teachers. Some homework assignments would help students make it a custom to learn regularly.

- Help students have full realization of the enjoyment of learning. Help students become more willing to learn. For example, make good connections between learning activities and the real world.

One of the approaches is introducing some additional teaching materials in the classroom, some of which are rather advanced materials for students who want to move ahead further based on mathematical ideas and skills they have learned. We call it advanced learning. Precisely speaking, since the first official Curriculum Standards were issued in 1958, this idea of advanced learning has been written into the standards. However, many people had not been aware of this idea and its importance.

The Ministry of Education released a resource book on mathematics education for schoolteachers titled *Examples of Advanced Learning and Complementary Learning in Mathematics Classrooms* in August 2002 (MEXT, 2002b). The books have been quite popular with schoolteachers who are interested in mathematics teaching.

SCHOOLING AND MATHEMATICS EDUCATION

The outline of Japan's system of school education is as follows:

- Kindergarten (under 6 years of age);
- Elementary school (1st through 6th grades);
- Lower secondary school (7th through 9th grades);
- Upper secondary school (10th through 12th grades); and
- University (postsecondary).

All children from the age of 6 to 11 are required to attend elementary school, and all children from 12 to 14 years of age are required to attend lower secondary school. The 6 years of elementary school education and the 3 years of lower secondary education are called compulsory education. Mathematics is a required subject at both of these levels. Teaching contents described in the Curriculum Standards are taught to all students. There are no exit tests at the elementary and the lower secondary school level. About 97% of lower secondary school students go on to upper secondary schools. Some students take competitive entrance exams, conducted by local governments or schools themselves, to enter popular upper secondary schools. The upper secondary school level includes both required and optional mathematics in the curriculum.

The Ministry of Education determines curriculum standards for schools from kindergarten to upper secondary school level so that a standardized education is available anywhere in Japan. The ministry has set the nationwide Curriculum Standards since 1947. The first several versions of the standards were provisional ones. Since 1958, the Curriculum Standards have been part of legislation by the Ministry of Education. For the elementary school level, since 1947 when the provisional Curriculum Standards were introduced, there have been six revisions (1951—provisional, 1958—first official version, 1968, 1977, 1989, and 1998).

Tables 2.1–Table 2.5 show the present status of teaching contents and total number of lesson hours for each subject. They are described in the standards or in related legislation.

TEXTBOOKS

Most textbooks are written by private sector publishers. The ministry approves them in accordance with the Curriculum Standards. All the contents described in the Curriculum Standards must be included in the textbooks. Some other advanced mathematics materials are included in the textbooks nowadays, because teachers and parents like to see them.

Local boards of education determine which authorized textbooks are adopted and used in the locality. Textbooks at compulsory education levels are offered to students free of charge. There are six publishers that produce mathematics textbook series for the elementary and lower secondary school levels.

Textbooks are the main materials for teaching and learning in classrooms. Some materials other than textbooks, such as exercise booklets, may be used in classrooms according to the school's judgment.

Figure 2.1 presents a flow chart of the process of textbook writing, authorization by the Ministry of Education, adoption, and so on. It takes about 3 to 4 years from the start of textbook writing to the use of them in the classrooms.

Table 2.1. Outline of Teaching Contents for Elementary School Mathematics

Grade	Numbers and Computations	Quantities and Measurements	Geometrical Figures	Mathematical Relations
1	• Meaning of integers • Addition and subtraction of integers	• Comparison of lengths	• Shapes around us	
2	• Addition and subtraction of integers • Multiplication table (9×9)	• Units of length • Units of time	• Shapes around us	
3	• Addition and subtraction of integers • Multiplication of integers • Division of integers • *Soroban*	• Units of length, volume, and weight	• Box shape • Squares, triangles, right triangles	• Table and bar graph
4	• Round numbers • Division of integers • Meaning of decimals • Addition and subtraction of decimals • Meaning of fractions	• Units of area • Area of squares and rectangles • Units of angle	• Isosceles triangles, regular triangles • Angles • Circles	• Table and graph of broken lines • Classification and organization of data
5	• Multiplication and division of decimals • Addition and subtraction of fractions • Estimation	• Area of triangles and parallelograms • Area of circles	• Parallelograms, trapezoids, rhombus	• Percent • Circular graph, band graph • Basic idea of functions
6	• Divisors and multiples • Addition and subtraction of fractions • Multiplication and division of fractions • Estimation	• Units of volume • Volume of cubes and rectangular parallelepipeds • Units of speed	• Cubes and rectangular parallelepipeds • Prisms, circular cylinder	• Ratio • Graph of proportion • Average

Table 2.2.   Outline of Teaching Contents for Lower Secondary School Mathematics

Grade	Numbers and Algebra	Geometry	Mathematical Relations
7	• Positive and negative numbers • Algebraic expressions • Linear equations	• Drawing of figures • Solid figures	• Direct proportion, reciprocate proportion
8	• Algebraic expressions • Simultaneous equations	• Plane figures • Proof • Congruence of triangles	• Linear functions • Probability
9	• Square root • Expansion and factorization • Quadratic equations	• Similarity of figures • Pythagorean Theorem	• Functions $(e.g., y = ax^2)$

Table 2.3.   Outline of Teaching Contents for Upper Secondary School Mathematics

Every student is required to take either Mathematics 1 or Basic Mathematics[*]

Mathematics 1	Basic Mathematics
• Equation and inequality • Quadratic function • Geometry and measurement	• Mathematics and human activities • Mathematical considerations in our society • Everyday statistics

The following courses form two optional sequences

Mathematics 2	Mathematics A
• Equations of higher degree • Geometry and equations • Logarithmic functions, exponential functions • Basic calculus	• Plane geometry • Sets and logic • Probability

Mathematics 3	Mathematics B
• Limit • Differential calculus • Integral calculus	• Number sequences • Vectors • Statistics and computers

	Mathematics C
	• Matrices • Curves and algebraic expressions • Probability

Note: *Most students take Mathematics 1.

Chapter 1—General Guidelines
Chapter 2—Teaching subjects
    1—Japanese Language
    2—Social Studies
    3—Mathematics
    4—Science
    5—Music
    6—Art
    7—Physical Education
    8—Technology and Homemaking
    9—Foreign Language
Chapter 3—Moral Education
Chapter 4—Special Activities

Mathematics is located in chapter 2 at both the elementary level and the secondary level. The mathematics part in the standards is composed of the following three parts, with the largest section being the teaching contents.

1. Objectives of teaching mathematics;
2. Teaching contents for each grade; and
3. Remarks on making teaching plans.

The next four elements are included in the objectives of teaching mathematics:

- Knowledge;
- Skills;
- Ability to think mathematically; and
- Interest in mathematics, willingness to learn mathematics, attitudes toward mathematics.

The following ideas are stated in the objectives of teaching mathematics:

- Learn mathematical knowledge and skills with rich sense of number, quantity, and geometrical figures;
- Think logically in good perspectives;
- Find the pleasure of doing mathematical activities; and
- Make good use of mathematics in daily life situations.

Teaching contents are divided into four areas at the elementary school level:

- Numbers and Computations;
- Quantities and Measurements;
- Geometrical Figures; and
- Mathematical Relations.

Throughout the Numbers and Computations section, the following three points are emphasized in the latest version of Curriculum Standards:

- Understand the meaning of computations;
- Consider computational methods; and
- Get good skills and make use of them.

Teaching contents are divided into three areas at the lower secondary school level:

- Numbers and Algebra;
- Geometry; and
- Mathematical Relations.

The following ideas are mentioned in the "Remarks on Making Teaching Plans." These ideas should be respected by teachers in mathematics classrooms:

- Learn by doing mathematical activities;
- Develop logical reasoning and intuition;
- Develop problem-solving abilities;
- Connect mathematics with daily-life situations;
- Use estimation when considering the process or checking results;
- Enrich sense of numbers, quantities, and geometrical figures; and
- Make use of *soroban* (abacus), calculators, and computers when solving problems.

Basically speaking, the Curriculum Standards are to show objectives and teaching contents. The main part of the standards is the teaching contents. Teaching methods and lesson plans should be devised by schools and schoolteachers. For example, the use of calculators and computers in mathematics classrooms is up to the school and its teachers.

As mentioned earlier, the Curriculum Standards have a characteristic of minimal standards. All the teaching contents mentioned in the standards must be taught in all the schools. Any material not mentioned in the standards may be added in each school curriculum as advanced learning material or as supplementary material, based on the needs of each student. A recent ministry survey shows that, after the implementation of the latest standards, more schoolteachers are teaching advanced materials in mathematics classrooms.

THE PROCESS OF REVISION OF THE CURRICULUM STANDARDS

The process of the recent curriculum reform can be summarized as follows:

1989 March	Release of the sixth standards
1992 April	Implementation of the sixth standards
1993-1995	Nationwide survey on students' achievement
1996 August	Report of the School Curriculum Committee
1998 July	Report of the School Curriculum Committee
1998 December	Release of the seventh standards
2002 February	Nationwide survey on students' achievement
2002 April	Implementation of the seventh standards and implementation of complete 5-day school week system
2004 February	Nationwide survey on students' achievement
2005 March	Education minister's consultation to the Central Education Committee on how to improve the Curriculum Standards

The Central Education Committee has been organized as part of the Ministry of Education. This committee serves as the education minister's consultation organization. The School Curriculum Committee is a subcommittee of the Central Education Committee. The basic philosophy of the revision of the Curriculum Standards is to be discussed in the School Curriculum Committee. The School Curriculum Committee has several subcommittees. A special mathematics education committee is one of them.

The members of the special committee on mathematics education so far have been chosen, for example, from college faculty (both mathematicians and mathematics educators), schoolteachers, local government education board officials, businesspeople working as computer engineers or architects, and representatives from the parents association.

Discussion in the committees is open to the public in principle. Committees accept opinion papers from academic circles, teachers' unions, parents' associations, labor unions, and so on.

For the discussion in the committees, the Ministry of Education conducts nationwide tests in mathematics, Japanese language, science, social study, and foreign language. The purpose of the tests is to gather data on the effectiveness of the Curriculum Standards. The data are supposed to be used in the process of curriculum reform. The latest national tests in mathematics were conducted in 2004. About 10% of students from fifth through ninth grade participated in the tests for this survey. These students are chosen randomly. Only nationwide results are supposed to be announced to the public, implying that local results are not announced.

The special committee makes the reports about their ideas on mathematics education and the next curriculum revision, and submits the reports to the School Curriculum Committee. The School Curriculum Committee and special committees exchange their opinions and negotiate to make final reports. The final report of the School Curriculum Committee is submitted to the Central Education Committee, and then to the education minister.

The Ministry of Education determines the Curriculum Standards according to the final report of the committee.

DEBATE ABOUT MATHEMATICS CURRICULUM

The latest version of the Curriculum Standards was released in 1998 and implemented in schools in 2002. There was a big change in the latest Curriculum Standards compared to the previous one. The total number of lesson hours for mathematics was reduced because of the introduction of the 5-day school week system, as well as the introduction of the new teaching subject called "integrated learning" from the elementary school level to the secondary school level. Mathematics lesson hours for elementary schoolers (first through sixth grade) were reduced from 1,011 to 869, and for lower secondary schoolers (seventh through ninth grade) from 385 to 315.

Additionally, one of the basic philosophies of the curriculum reform was that students should learn less content more expertly in the classroom. There was debate about the reduction to the mathematics curriculum in academic circles. Mathematicians and mathematics educators were generally against the reduction of lesson hours and teaching contents. However, many schoolteachers, parents, and people in society seemed not to be strongly against these reductions in the recent Curriculum Standards. Before the latest revision appeared in 1998, opinions such as "school mathematics is too difficult and too heavy for a lot of students, which is due to the Curriculum Standards!" were sometimes seen on the op-ed page of newspapers. But there were some schoolteachers who emphasized the importance of mathematics and were against the reduction of mathematics lesson hours and teaching contents.

As mentioned earlier, immediately after the latest revision, people in general said they were satisfied with the new Curriculum Standards. However opposing opinions gradually began appearing in newspapers and in other mass media. Some people charged that teaching fewer mathematics topics to students leaves them less knowledgeable and thus with weaker skills. This induced some parental anxiety about the education system.

These days, more people say that greater emphasis should be placed on mathematics teaching in schools. Some people say they are afraid that students' mathematics achievements might decline in the near future.

In March 2005, the education minister asked the Central Education Committee to discuss improving the Curriculum Standards. The minister's concern includes the improvement of mathematics and science education. The review of the teaching hours and teaching contents in mathematics should be one of the discussion themes in these sessions of the School Curriculum Committee.

Several discussion themes will appear in the School Curriculum Committee and in the mathematics education special subcommittee. We expect, for example, the following discussion themes to appear:

- For what reason is mathematics placed in school curriculum? What relations does mathematics have with other subjects? What portion of the total teaching hours in school curriculum should be assigned to mathematics education? Why?

- Mathematics contents prescribed in the Curriculum Standards are to be taught to all the students at elementary and lower secondary level. What mathematics contents are necessary to all the students? Why?

- What mathematics contents are necessary to live effectively in society? What mathematics contents are necessary to continue studying mathematics and other academic subjects further?

- Public support is needed to improve school mathematics education. We expect parents and people in society to get more interested in mathematics education and to encourage students learning mathematics. We want parents and people in society to root for schools, too. We expect teachers to get more interested in mathematics education, especially elementary school teachers, who teach all school subjects. What appropriate information on mathematics education should be presented to teachers, parents and the people in society?

- These days there are more people getting interested in students' achievements in mathematics. They expect schools to develop students' academic ability. How should schools respond to peoples' expectations?

- The ministry of education is planning to conduct a new type of national test in 2007. All sixth and ninth grade students will be able to join the test on mathematics and Japanese language in accordance with the Curriculum Standards. The purpose of the new test is to help schools review their school curricula as well as their teaching methods, and also to help students develop the motivation to study. In what way should the results of the new test be announced?

REFERENCES

Ministry of Education, Culture, Sports, Science, and Technology. (1998a). *Yochien Kyoiku Yoryo* [Curriculum standards for kindergarten]. Tokyo: Printing Bureau.

Ministry of Education, Culture, Sports, Science, and Technology. (1998b). *Syogakko Gayusyu Sido Yoryo* [Curriculum standards for elementary school]. Tokyo: Printing Bureau.

Ministry of Education, Culture, Sports, Science, and Technology. (1998c). *Chugakko Gakusyu Sido Yoryo* [Curriculum standards for junior high school]. Tokyo: Printing Bureau.

Ministry of Education, Culture, Sports, Science, and Technology. (1999a). *Kotogakko Gakusyu Sido Yoryo* [Curriculum standards for High School]. Tokyo: Printing Bureau.

Ministry of Education, Culture, Sports, Science, and Technology. (1999b). *Syogakko Gakusyu Sido Yoryo Kaisetsu*, Sansu Hen [Guidebook of curriculum standards for elementary school mathematics]. Tokyo: Toyokan Syuppan.

Ministry of Education, Culture, Sports, Science, and Technology. (1999c). *Chugakko Gakusyu Sido Yoryo Kaisetsu, Sugaku Hen* [Guidebook of curriculum standards for junior high school mathematics]. Osaka, Japan: Osaka Syoseki.

Ministry of Education, Culture, Sports, Science, and Technology. (1999d). *Kotogakko Gakkusyu Sido Yoryo Kaisetsu, Sugaku Hen* [Guidebook of curriculum standards for high school mathematics]. Tokyo: Jikkyo Shuppan.

Ministry of Education, Culture, Sports, Science, and Technology. (2002a). *Manabi no Susume* [Encouragement of learning]. Tokyo: Author.

Ministry of Education, Culture, Sports, Science, and Technology. (2002b). *Hattenteki na Gakusyu ya Hojuteki na Gakusyu no Suishin, Shogakko Sansu* [Examples of advanced learning and complementary learning in mathematics classrooms]. Tokyo: Kyoiku Syuppan.

CHAPTER 3

AN OVERVIEW OF MATHEMATICS EDUCATION IN SINGAPORE

Cheow Kian Soh
Ministry of Education, Singapore

INTRODUCTION

The education system in Singapore has undergone many changes since the early 1960s. These changes mirror the development of the country, the priorities of the education system, and the aspirations of the people. This chapter gives an overview of mathematics education—the review and development process, the support for implementation, the assessments and examinations and finally, the challenges.

CHANGING LANDSCAPE

Singapore gained independence in 1965. In the early days of nation building, the priority was to give every child a place in school. While there were alternative syllabi at the higher grades, there was little attempt to further differentiate the mathematics curriculum. A uniform mathematics curriculum was available to most students.

Mathematics Curriculum in Pacific Rim Coutries—China, Japan, Korea, and Singapore: Proceedings of a Conference, pp. 23–36

In the 1980s, the economy was largely driven by manufacturing. The skill and productivity of the labor force were important. The education system geared itself to providing all students with at least 10 years of general education. By then, mathematics was compulsory in secondary schools. Streaming by ability was introduced. Different curricula for the different streams and courses were designed to meet the varying needs and abilities of the students. It was the first systemic attempt at curriculum customization.

The key driving forces in the late 1990s were globalization, the needs of a knowledge-based economy, and information technology. The ability to create knowledge, innovate, and adapt to a changing environment was important. There was greater focus on harnessing the talents and abilities of every individual. In response to these new challenges, three initiatives were launched, National Education, Thinking Skills, and Use of Information Technology. The buzzwords were "ability-driven education" and "mass customization."

Going forward, to meet the needs and aspirations of the people, more opportunities and greater choice and flexibility in the education have now been introduced. New types of schools and a wider range of curricular options have been made available. Specialized independent schools with their own niches and programs have been set up. The new Integrated Program which offers seamless education from secondary to preuniversity is now offered to cater to more able students. Besides the general certificate of education (GCE) examinations, an international baccalaureate type program is also available to students.

At the heart of all these changes is the mission to provide the best education for the students. The call to "Teach Less, Learn More" focuses on the quality of teaching and learning, quality of interaction between teachers and students, and quality of school experience as a whole. It is a call for educators to better engage learners and to prepare them well for life. "Teach Less, Learn More" is about a qualitative change, not a quantitative cut.

EDUCATION STRUCTURE TODAY

The education structure is generally a 6-4-2 system, with 6 years of primary school education (Grades 1-6), 4 years of secondary (Grades 7-10) and 2 years of preuniversity (Grades 11-12). Students move on to schools at a higher level at the end of each stage. There are, however, a few full schools that offer both primary and secondary education or both secondary and preuniversity education at the same campus. Figure 3.1 shows the education and progression structure of the Singapore education system.

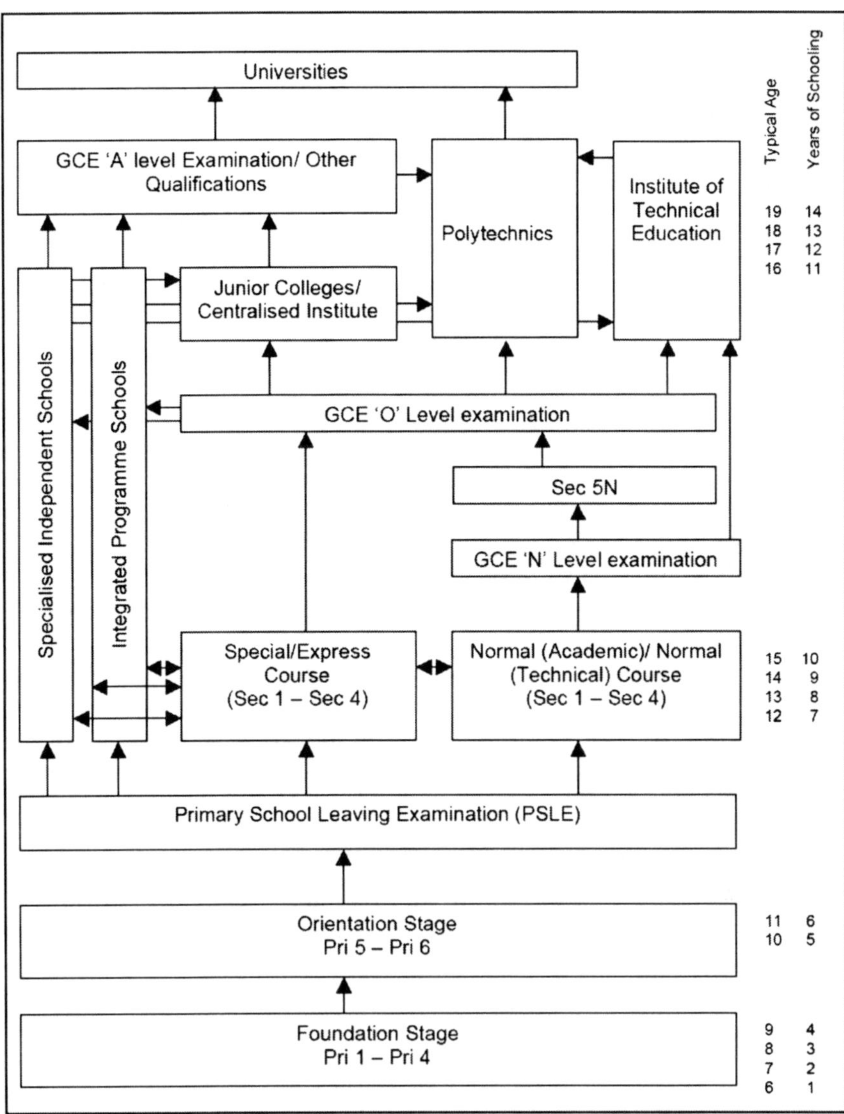

Figure 3.1.   Education and progress structure.

DIFFERENT NEEDS, DIFFERENT PATHWAYS

Primary education consists of two stages—a 4-year foundation stage (Grades 1-4) followed by a 2-year orientation stage (Grades 5-6). The curriculum during the foundation years is largely common. At the end of Primary 4 (Grade 4), students who would benefit from a slower pace of learning are identified by their schools. These students are offered an "EM3" curriculum that better caters to their needs and abilities. All students eventually sit for the Primary School Leaving Examinations (PSLE) at the end of Primary 6 (Grade 6). However, the students in the EM3 take different exams.

There are three courses at the secondary levels. Students are placed in the Special/Express,[1] Normal (Academic), or Normal (Technical) courses based on their performance at the PSLE. The 4-year Express course leads to the GCE O-level certification. These students typically move on to a preuniversity course or a polytechnic education. The Normal (Academic) course can be seen as a 4-year course leading to the GCE N-level certification or a 5-year course leading to the O-level certification. The 4-year Normal (Technical) course is geared toward preparation for technical- and service-oriented courses at the Institutes of Technical Education. It also leads to the GCE N-level certification.

Preuniversity education prepares students who wish to pursue a course of study at the university upon graduation. The junior colleges offer a 2-year program leading to the GCE A-level certification while the sole centralized institute offers a 3-year program.

On average, about 20-25% of each cohort of students eventually proceeds to the universities. About 60% of each cohort of students are in the Express course, 25% in the Normal (Academic) course, and 15% in the Normal (Technical) course.

FLEXIBILITY AND CHOICE

There is flexibility and choice for students within this main structure. There are also possibilities of transfers across streams and courses, giving opportunities for students to be placed in a course for which they are most suited. Students who performed well in the Normal (Academic) course could choose to study in the Express course; those in the Normal (Technical) course have the same flexibility to study in the Normal (Academic) if they perform well at the examinations. The system also allows for the more able students from the Normal (Academic) course to sit for the O-level examination for some subjects at the end of the fourth year instead of the fifth year.

KEY PLACEMENT EXAMINATIONS

The PSLE, GCE O-Level, N-Level Examinations, and A-Level Examinations are key examinations for the students. Not only do these examinations give an indication of the achievement levels of the students but they also play an important role in deciding whether a student would be posted to his or her preferred choice of school or be selected for special programs. They are therefore viewed as high-stakes examinations by students, teachers and parents.

In recent years, we have allowed selected secondary schools and junior colleges to admit students outside the central posting system.[2] This new scheme allows schools with niche academic or nonacademic programs to set aside a percentage of the places for deserving students who are gifted and talented in the niche areas of the schools. The students can gain entry without relying solely on the outcomes of the key examinations. This move not only aims to recognize special talents but it broadens the definition of success beyond just marks and grades. It also hopes to relieve some of the stress related to the key examinations.

MATHEMATICS EDUCATION

Mathematics education begins early. From the time a child enters the formal school system at Primary 1, about age 6 or 7, mathematics is part of the main diet. Mathematics remains a compulsory part of the school curriculum up to the end of secondary education. This gives every child about 10 years of education in mathematics.

The mathematics curriculum is centrally planned, with flexibility in the implementation at school level. We believe that a large part of what a child needs to learn in mathematics in the formative years is common and requires careful thought and planning to make it accessible to every student. A centrally planned curriculum provides clear guidance in teaching and learning to teachers. Flexibility is given to schools in how they implement the curriculum so as to best meet the needs and abilities of their students.

POLICY OBJECTIVES

Mathematics education is an important part of the national curriculum. It contributes to the development of the individual as well as in meeting the needs of the nation. For the individual, an understanding of mathematics provides the concepts and skills to understand the world around one, to function effectively at work, life, and play and to support the learning of other disciplines. For the nation, a strong grounding in mathematics provides the basis and support toward building a highly skilled and well-educated workforce both scientifically and technologically, which in turn promotes a competitive and healthy economy and society.

The policy objectives for mathematics education are twofold. First, the mathematics curriculum aims to provide all students with a firm foundation in mathematical concepts and skills that underpin a wide range of daily activities and uses. Second, it aims to provide students who have the aptitude and interest in mathematics the opportunities to deepen their knowledge and skills, and to pursue their passion in mathematics so that they will, in turn, contribute to the progress of the nation.

It is clear from the policy objectives that a one-size-fits-all mathematics curriculum will not be able to serve the needs of all students. A differentiated and targeted approach, with choice and flexibility, is part of the overall strategy to achieve the objectives of mathematics education in Singapore.

MATHEMATICS CURRICULUM FRAMEWORK

A single curriculum framework is used consistently throughout the different levels, differing only in the details at each level but sharing common emphasis throughout the levels (Ministry of Education: Curriculum Planning and Development Division, 2006a, 2006b). The framework provides a high level perspective of the important components of the mathematics curriculum and serves as a framework to guide the implementation of an effective mathematics program in schools. Teachers are familiar with the framework and it has become a convenient tool to communicate and discuss the mathematics curriculum.

At the center of the framework is the focus of the mathematics curriculum: mathematical problem solving refers to the intelligent and creative use of mathematics as a means for solving problems. The use of mathematics as a tool and language to define, pose, formulate, solve and evaluate a problem is emphasized in different shades and hues within the mathematics curriculum.

There are five interrelated components that support the attainment of mathematical problem-solving skills. Students need to acquire a wide range of mathematical concepts and skills, and demonstrate understanding and proficiency in these areas. Generic processes, such as reasoning, communication, making connections, heuristics and thinking skills are also part of the learning in mathematics. Beyond that, meta-cognitive insight and control of one's thought processes are important skills required of a good problem solver. The affective domain should not be neglected. Cultivating a positive attitude towards mathematics will engender interest and perseverance in learning and doing mathematics.

The mathematics curriculum framework has been updated to reflect changing emphases and needs. Details have been added to the frame-

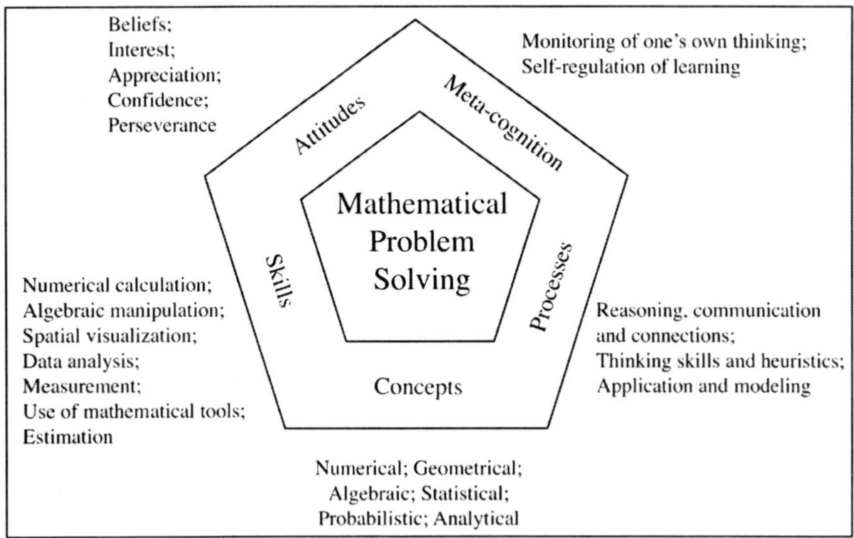

Figure 3.2.   2003 mathematics curriculum framework.

work since it was first developed in the 1980s. The framework was last updated in 2003. For example, in 2001 (Ministry of Education: Curriculum Planning and Development Division, 2001a, 2001b), the framework included thinking skills as part of the new initiative mentioned earlier. In 2003, responding to global trends and emphasis on knowledge skills, the importance of communication and making connections were made more prominent in the framework. The latest framework is given in Figure 3.2.

DIFFERENTIATED MATHEMATICS CURRICULUM

Primary (Grades 1-6)

In the foundation years, there is a common mathematics syllabus emphasizing basic number concepts, the use of numbers in measurements, understanding of shapes and simple data analysis. Problem solving involving meaningful contexts is part and parcel of the curriculum. Students learn through concrete manipulatives and simple diagrams and pictures. Drawing models to represent a problem and using the pictorial representations to see relationships and formulate strategies for solving problems feature prominently in the approach. Algebra, in a more formal sense, is introduced late in the elementary years. However, algebraic concepts such as unitary thinking, and proportional thinking, are evident in the problems that students are required to solve.

At the end of Primary 4, students are streamed according to their ability. The mathematics curriculum for the EM3 repeats and reinforces some of the early concepts and skills learned during the foundation stage. The pace is slower and the content is reduced. For example, decimals are only taught later while ratio, speed and algebra are postponed to the secondary level for these students. This is the first point of differentiation of the mathematics curriculum.

Secondary (Grades 7 to 10)

Students continue to learn mathematics throughout the secondary school years. This means that all students would have up to 10 years of mathematics education. However, not all students are expected to learn the same amount of mathematics; much depends on the ability and interest of the student. There is greater differentiation within the curriculum at the secondary levels.

There are three different mathematics syllabi catering to the students in the three courses. In addition to the main mathematics syllabi, students from the Express and Normal (Academic) courses who are interested and more mathematically inclined are offered an additional mathematics subject, for greater breadth and depth. This additional mathematics subject is offered from Secondary 3 (Grade 9) and aims to prepare students for further study in mathematics or disciplines where more mathematics is required. The additional subject includes introductory topics in calculus. Close to 60% of the secondary students opt for the additional mathematics course.

The mathematics syllabus for the Normal (Academic) course is designed as a subset of the syllabus for the Express Course. It is designed with the intention to provide a slower pace of learning that could dovetail to the preparation for the GCE O-Level in the fifth year. The focus of the mathematics syllabus for the Normal (Technical) course is functional. It emphasizes the use of mathematics in everyday applications and prepares students for a technical or service oriented course at the technical institutes.

All in all, there are five different syllabi for the three secondary courses—a different mathematics syllabus for each of the three courses and two additional mathematics syllabi for the Express and Normal (Academic) courses. As mentioned earlier, there is flexibility for the more able students from the Normal (Academic) to take the mathematics syllabus in the Express course at the end of the fourth year instead of the fifth, and for the Normal (Technical) to take the mathematics syllabus for the Normal (Academic) course. This variation in the pace of learning adds to the options available to students.

Preuniversity (Grades 11 to 12)

At the preuniversity level, mathematics is not a compulsory subject. Currently, about 90% of the students take some form of mathematics at this level. Three options, H1, H2, and H3 mathematics, are offered to the students.

The H1 syllabus provides a foundation in mathematics for students who intend to enroll in university courses such as business, economics and social sciences. The H2 syllabus prepares students for university courses such as mathematics, physics, engineering and computing where more mathematics content is required. Students will learn to analyze, formulate and solve different types of problems. Statistics features prominently in both syllabi and the use of graphing calculators is expected. For students who have a strong aptitude for, and are passionate about mathematics, the H3 syllabus offers an opportunity to further develop their mathematical modeling and reasoning skills.

REVIEW OF THE MATHEMATICS CURRICULUM

The review of the mathematics curriculum moves in tandem with the major reviews, repositioning and meeting changes in the education system. On average, a major syllabus review is conducted every six years, with a mid-term syllabus review at the third year of implementation. The frequency of the review also depends on factors, such as trends and developments in mathematics education around the world and the feedback from schools on implementation of the syllabi.

KEY CONSIDERATIONS

In developing the syllabi, there are several key considerations that shape syllabus content and approaches. Among these considerations is that the content and approaches should be appropriate for the level and course, with applications that are relevant and meaningful to the targeted group. As the different syllabi serve different groups of students, the choice of content and approaches varies. For example, the Normal (Technical) mathematics syllabus has relatively less algebra and more computations, applications are less abstract with more day-to-day contexts, and there are fewer theoretical tasks while there are more hands-on activities. At the primary level, the use of a concrete-pictorial-abstract approach that is aligned with the development stages of the child is emphasized in developing the sequence of topics and materials to support the learning.

With the many syllabi interacting and connecting with each other, there is a need to maintain the coherence and continuity within and across levels and syllabi, both vertically and horizontally. Care is taken to ensure continuity at the design stage so that students can move from one level to the next smoothly. Assumptions have to be made about how students would progress. For example, in developing the syllabus for the Normal (Technical) course, the assumption that students are mainly from the EM3 curriculum means that the starting point for the secondary syllabus has to be lower compared to the syllabus for the Express course.

The learning outcomes of the syllabi are organized along strands, such as numbers and algebra, geometry and measurement, and probability and statistics. However, in practice, mathematics is taught as a subject without strand boundaries. Connections across strands and applications involving concepts and skills from more than one strand are encouraged.

A spiral approach, with one layer of content built on the next and the revisiting at greater depth at the next level, is an integral part of the design. The grade-by-grade development sequence is most prominent at the primary level, where the learning outcomes at each grade level are specified. However, at the upper secondary level and beyond (Grades 9-12), learning outcomes are grouped in bands of two grade levels for greater flexibility in teaching and learning.

There is also the consideration of time to ensure that the curriculum is effectively delivered. Over the years, there has been a careful reduction of content so that time and space could be made available to teachers to teach the content and skills, and for students to learn the content well. Maintaining rigor, while reducing quantity, requires careful consideration. Much thought is given to the identifications of core skills and concepts that would support future learning.

REVIEW COMMITTEE

In the most recent review started in 2003, a single committee led by the Ministry of Education (MOE) was formed to review all the mathematics syllabi in the curriculum, from the primary through the secondary to the preuniversity levels. This was a break from past practice, where smaller review teams were set up for each syllabus. Having a single review committee was part of the effort to ensure better alignment, coherence and continuity among the mathematics syllabi and across the levels.

The Mathematics Syllabus Committee (MSC) is comprised of curriculum specialists and curriculum planning officers, teachers from schools (primary to preuniversity), mathematicians and mathematics educators from tertiary institutions, and representatives from the Singapore Examination and Assessment Board (SEAB). The tertiary institutions include the local universities, the polytechnics and the institutes of technical education.

CONSULTATION WITH STAKEHOLDERS

Much of the work during the review stage involves consultation with teachers and other stakeholders. The curriculum specialists and curriculum planning officers (who are teachers serving a stint at MOE) from the Mathematics Unit are actively involved in the work of the MSC. Focus group discussions with teachers at different stages of the review are held and the findings are shared and deliberated at the committee meetings. Draft syllabi are planned and reviewed based on feedback from the schools.

Feedback from teachers is important to the planning process. Adjustments of topics across grade levels and resequencing of topics are made based on best practices identified in schools. The process is iterative, with practice informing design and design influencing practice. As there is no perfect sequencing or design in syllabus implementation, the syllabi serve as a guide and teachers have the flexibility to interpret and implement them in the best way for their students.

At the secondary and preuniversity levels, the examiners from the Cambridge International Examinations are also consulted. This is necessary to ensure that the assessment objectives of each syllabus are aligned to the intent and scope of the syllabi. For example, when reviewing the mathematics syllabi for the 2006 A-Level curriculum, the expectation that students would use a graphing calculator during the examination led to several rounds of syllabus adjustments and changes to the specimen papers.

SUPPORT FOR IMPLEMENTATION

For successful implementation, communication with and support for teachers are two important areas of work carried out and coordinated by the Mathematics Unit. The National Institute of Education (NIE), which takes charge of preservice training for teachers, plays a key partnering role in providing in-service training and developmental opportunities for our mathematics teachers.

COMMUNICATIONS WITH MATHEMATICS TEACHERS

To maintain a two-way communication between MOE and the schools, the Mathematics Unit holds meetings with the heads of department to disseminate, discuss and share information at least twice a year. The meetings provide a forum and opportunity to update the teachers of ongoing reviews, discuss areas of concern and share best practices. In addition to the meetings, curriculum planners also visit schools and hold focus group discussions to gain a better understanding of the ground. All these provide useful information for policy formulation and decision making.

TRAINING AND SHARING FOR MATHEMATICS TEACHERS

Teachers are keys to the successful teaching and learning of mathematics. All teachers receive their preservice training at NIE. The training covers pedagogy as well as content areas. There are also in-service courses run by MOE, NIE, and other agencies. The universities and polytechnics also provide training in niche areas in mathematics.

There are ample opportunities for teachers to share and learn from each other. The culture of sharing, which has gathered speed and momentum since the late 1990s, is one of the strategies used to constantly upgrade our teachers professionally and to ensure that they are well equipped with the knowledge and skills to deliver the curriculum effectively.

MATHEMATICS TEXTBOOKS AND RESOURCES

New textbooks are produced whenever there are changes in the syllabi. The textbooks translate the emphases, approaches and scope of the syllabi into concrete forms for teachers to use. They serve as essential learning materials for teachers to deliver the syllabi and for students to learn independently.

Since the late 1990s, MOE has devolved textbook writing and production to commercial publishers, not just for mathematics but for most of the subjects. Instead of a single textbook for all schools, the participation of publishers allows for a greater variety of textbooks that would meet the learning abilities of students and yet which are aligned with the spirit and intent of the syllabi. Quality is assured through a rigorous textbook authorization and approval process involving textbook reviewers who are teachers, curriculum planning officers and academic resource persons. As the students are required to buy their textbooks, affordability of the instructional materials is also an important consideration.

Besides the textbooks, there are other resources that are available to teachers. Some of these resources, such as the teachers' guide, are provided by publishers. The Mathematics Unit also produces in-house materials to support schools in specific areas. Teachers are also encouraged to create their own resources. This is particularly so for the preuniversity level, where there are generally no common textbooks written specifically for schools.

MATHEMATICS ASSESSMENTS AND EXAMINATIONS

Teachers regularly conduct assessments to inform them of their students' progress and achievements in learning. Typically, assessments and tests are carried out at the end of each term, with a mid-year and an end-of-year examination. Schools can decide on the format and frequency of the tests. MOE provides guidelines to help schools in making decisions about assessment.

Other assessment methods such as the use of reflective journals and project-based tasks are gaining greater acceptance. MOE has been encouraging teachers to widen their repertoire of assessments so that they are able to assess not only the understanding of concepts and skills, but also the other aspects of the mathematics framework such as the affective domain, the process component and meta-cognitive awareness.

A more traditional pen-and-paper test is used in summary assessments in schools, similar to those used in the national examinations. Typically, questions focus on concepts and skills, including problems set in a variety of contexts and applications.

CHALLENGES

The Singapore education system has put in place a flexible education structure to cater to the different needs and abilities of the students. However, there is a limit to what systemic and structural changes can achieve. Going forward, we would like schools to further customize their mathematics program and for teachers to put emphasis on differentiation in classroom instruction. In this way, we would better engage students and give every one of them the best opportunity to learn mathematics.

There is more that we can do for students at the extreme ends of the ability spectrum. For students who have difficulty learning mathematics, we will continue to find ways to support their learning. For students who are gifted, we will continue to provide them with more opportunities to achieve higher peaks of excellence. For all students, we will continue to make the learning of mathematics relevant, meaningful and fun.

There is a tendency to focus too much attention on preparing our students for assessment and examination. While examinations are an important part of the education process, providing motivation and signposts for achievements, an overemphasis on examination results could compromise the quality of learning. Using assessment for learning, and not learning for assessment, is one of the reasons for encouraging schools to use a wider range of assessment strategies to inform teaching and improve learning.

CONCLUSION

The education landscape is changing. Some of the changes described in this paper are new and some are in the pipeline. The latest review of the mathematics curriculum was completed in 2004 and will be implemented in stages from 2006 onward. The mathematics curriculum will continue to evolve, taking in the best practices from our schools and around the world. Every system has its own unique strengths, constraints, and challenges thus each will have to find its own unique solutions.

NOTES

1. Henceforth, I will refer to this as the Express course as the difference between the Special course and Express course is only in the learning level of the mother tongue.
2. The central posting system is based on the performance of students at the key examinations and the students' choice of schools.

REFERENCES

Ministry of Education: Curriculum Planning and Development Division. (2001a). *Primary mathematics syllabus*. Singapore: Author.
———. (2001b). *Secondary mathematics syllabus*. Singapore: Author.
———. (2006a). *Mathematics syllabus primary*. Singapore: Author.
———. (2006b). *Secondary mathematics syllabuses*. Singapore: Author.

CHAPTER 4

SOME CHARACTERISTICS OF THE KOREAN NATIONAL CURRICULUM AND ITS REVISION PROCESS

Hee-Chan Lew
Korea National University of Education

Education is a key factor to understanding Korean society. Across Korean society there is substantial interest in children's education and Korean parents make great investments into their children's education with the prospect of providing their children with a better life. Most Korean parents believe that their children's education is valuable enough for them to endure real hardships (Lee, 2002). According to the Korean Educational Development Institute (Kim, Yang, Kim, & Lee, 2001), public education fees including college tuition in 2000 equaled 33.5 billion U.S. dollars, whereas outside-of-school tutoring expenditures were calculated at 37 billion U.S. dollars. Parents in Korea are spending approximately 10% of their income on their children's out of school tutoring expenses.

Such keen interest in education is observed as one of the main reasons why Korean students fare so well in such international comparative studies as Trends in International Mathematics and Science Study (TIMSS),

Mathematics Curriculum in Pacific Rim Coutries—China, Japan, Korea, and Singapore:
Proceedings of a Conference, pp. 37–71

Third International Mathematics and Science Study-Repeat (TIMSS-R), or Program for International Student Assessment (PISA). Considering that aspiring teachers and educators (students enrolled in schools of education and teacher training programs) tend to display a considerably high achievement level compared to their peers in high school and that those who wish to pursue a career in teaching must pass the extremely competitive Teacher Employment Examination, teacher quality is a very strong variable in the Korean context (K. Park, 2004). It can also be seen that the national curriculum of Korea responded quite positively to such international comparative studies. However, it is also arguable that Korean students were able to score well in international comparative studies due to the extensive out-of-school tutoring, which exposes students to numerous and varied mathematical problems in preparation for their college entrance exam.

There have been many changes in Korea's mathematical education since 1997, when the TIMSS results were released and informed the world of Korean students' high achievement level in math. The 7th National Curriculum, which was stated as an attempt to revolutionize the national curriculum from the 6th, was promulgated in December 1997. There was a movement among teachers promoting voluntary teaching and improvement of assessment methods from the late 1990s into the beginning of the twenty-first century. Also, there was an active attempt to introduce computers into math classrooms. Such tides of change after the TIMSS results can be summarized as math education that advocates activity-oriented learning and enhances thinking skills.

The factors leading to such changes can be classified into three branches. First is the TIMSS study results (Beaton et al., 1996; Mullis et al., 1997). The analysis on the different types of math classrooms that was set forth by the TIMSS report had instilled a new found confidence in the hearts of math educators of Korea and also motivated them to sincerely reflect on their teaching methods, consequently constituting an opportunity to systematically look back on the math education of Korea. Second, the overall flow of change in math education can be considered. Since the revision of the 4th National Curriculum released in 1980, the teaching of math had slowly but largely shifted from theory-based mathematics to real-life-based mathematics. Such changes have their origin in an ongoing effort to nurture a qualified workforce that meets the needs of a modern developed economy. Spurring these considerations was a period of dramatic economic growth beginning in the 1960s. The third factor is the strong influence of American math education that started to emphasize problem solving and application since the 1980s. As Korean international students who had studied in the United States started to return to Korea in the late 1980s, the demanded methodology of the times, which called for improving problem-solving ability, was aggressively introduced into Korean education.

This paper has its purpose in stating the main contents of Korea's 7th National Curriculum which was revised after the release of the TIMSS results, identifying the factors that served toward such changes, and looking into what process revision of the curriculum had followed. With these aims this paper will first project the historical and societal relationship between Korea's social structure and education, and then introduce the diagnosis that was set forth by TIMSS on Korea's math education, ultimately influencing revision of the 7th National Curriculum.

THE RELATIONSHIPS BETWEEN KOREA'S SOCIAL STRUCTURE AND EDUCATION: THE CORRELATION BETWEEN TESTS AND UPWARD SOCIAL MOBILITY

To understand Korea's educational system requires a thorough knowledge of both the public education system, of which the national curriculum and textbook system are foundational, and the private-sector educational offerings. According to Kim et al. (2001), 84.1% of Korean parents provide some form of shadow education for their children (primary 91.1%, middle school 81.5%, high school 70.2%) and it has been reported that this phenomenon is becoming more prevalent.

Korean parents have a strong belief that their children's future is realized through education and that acceptance to a top university will guarantee bright prospects for their children's future. The rationale behind why Koreans have come to rely so much on private tutoring, rather than public education solely, lies in parental beliefs that children need to study more intensely and more broadly than their peers in order to be accepted by these top universities.

The image in Figure 4.1 is a poster publicizing a special exhibit on education, commemorating the 60th anniversary of Korea's liberation from Japan, that was held at Korea National University of Education in September 2005. A catch phrase on the poster, "Education was only our hope" announces the vigor with which Korean society approaches the idea of education.

Then for what reason are parents in Korea so engrossed in their children's education that they are willing to sacrifice a great deal on their own part? A first reason lies in Korean society's long-standing tradition of rewarding exam results with upward social mobility. The state examination (*Kwakeo*), which was first introduced in the *Koryo* era (A.D. 918-1392) in AD 958 from China, remained implemented for the next 950 years until the late *Choseon* era (A.D. 1392-1910). This exam was an administe-

Figure 4.1.   Poster for a special exhibit on education commemorating the 60th anniversary of Korea's liberation from Japan.

rial tool used to recruit talented young people (Woo, 1992). It also instilled a perception in the common people that being recruited as a country officer by virtue of the exam was an invaluable honor for both the family and the individual. This also has strong ties with the traditional perception of bureaucrats being more associated with prestige than their counterparts who were engaged in agriculture, industry and commerce in the *Koryo* era and the subsequent *Choseon* era. This was a time when the society was strictly stratified into three classes such that the lower classes were not allowed to apply for the state examination; only the upper classes were able to take the examination required for becoming a general or high-ranked officer, while the middle classes were only eligible to take the examination required for becoming skill-oriented mid-rank bureaucrats.

Under Japanese occupation for 35 years, from 1910 until 1945, the class system was eradicated. In the 20 years that preceded the Second

World War, there was regular and consistent chaos in the political and social systems of Korea (1945-1948 chaotic conflict, 1950 Korean War, 1960 Students Revolution, 1961 Military Revolution). Although the contemptuous social climate toward people engaged in agriculture, industry and commerce had mostly disappeared, there was still a common understanding that passing an exam could win one amiable treatment within the society.

Amidst such devotion toward private tutoring, math is placed at the center. Parents and students alike believe that other subjects are subjects in which scores can be significantly increased by rote memorization within a short amount of time. But, they do not believe that this is the case with math. Therefore 70-90% of students from lower graders through high school seniors participate in some form of private-sector educational institution for at least 2 hours everyday. However, a problem with these outside of school programs is that students learn technical skills for improving achievement scores in school and later for preparing for the college entrance examination rather than cultivate their mathematical ability. As such, with the help of private tutoring, students are readily able to solve problems provided in any type of exam, and as a result have achieved extremely high scores in such international contests as TIMSS. Due to this, there is concern about whether or not such high scores achieved in the TIMSS study accurately reflect the mathematical abilities of Korean students.

THE REALITY OF KOREAN MATHEMATICS EDUCATION REFLECTED THROUGH TIMSS

In the TIMSS, in which over 40 countries participated, Korea scored higher than all other countries except Singapore with scores of 607 and 611 among the eighth graders and the fourth graders respectively (Beaton et al., 1996; Mullis et al., 1997). Korea also has attained similarly positive results in other large-scale international competitions administrated in the last 2 decades, such as the International Assessment of Educational Progress (IAEP) I and II (Im & Kim, 1995; Educational Testing Service, 1989).

The Korean students' internationally recognized high achievement level strongly encourages many Korean mathematics educators who believe that their students' mathematical potential is very important with respect to national economic development and to strengthening its competitiveness in the international market place.

International studies, like TIMSS, have provided Korean mathematics educators with a matchless opportunity to reflect on their educational environment as a whole. Particularly, TIMSS showed that Korean mathematics education has many serious weak points. Korean mathematics educators take precious lessons from TIMSS for improving their current situation.

First, Korean students' affective characteristics were not friendly toward mathematics compared with other countries. In the case of the eighth grade, Korea was one of the lowest countries along with Japan and Lithuania in terms of confidence and interest levels. Sixty-two percent of the students disagreed with the statement that it is important to do well in mathematics and 42% of them disliked mathematics. As well, 50% of the students expressed an overall negative attitude towards mathematics. What is more serious is that this phenomenon is becoming even more commonplace in the higher grades. In the case of the fourth grade, the corresponding rates indicating interest and attitudes were 27% and 28% respectively. In addition, 53% of Korean students disagreed that the reason why they need to do well in mathematics is get a desirable job. TIMSS results show that Korea's school system very urgently needs to develop a particular program for nurturing students' affective dispositions.

Second, there was a significant difference in the achievement levels between male and female, and between urban and rural areas. This result poses a serious problem with respect to equal opportunities in mathematics education. The androcentric traditions and customs usually guide girls, regardless of their mathematical talents, to choose college majors that are unrelated to mathematical fields. Parents and even teachers do not generally expect girls to fare as well in mathematics as their male counterparts. Furthermore, the educational environment in rural areas is generally inferior compared to urban areas in terms of teacher quality, parents' educational expectation levels, access to educational information, and so forth. Any modern information society cannot endure this kind of gap in that mathematics knowledge is needed for all social groups and communities in order to move forward.

Third, TIMSS results show that in Korea, whole class activities run under direct teacher control as the first consideration in classroom management. In the case of the eighth grade, 89% of the students, which was the largest rate among all TIMSS countries, reported that mathematics classes were run by working together as a whole with the teacher teaching in the front, and only 12% of students, one of the lowest rates, worked in pairs or small groups with assistance from teachers. These practices of whole class lessons that inevitably rely more on verbal explanations by teachers than students' spontaneous construction should be changed.

Fourth, according to the TIMSS report, Korean math teachers were the most conservative in using technology, such as the use of computers and calculators. The rate of teachers and students using calculators and computers in their classes and problem-solving processes were the lowest among countries participating in TIMSS. In the case of the eighth grade (fourth grade), 76% (86%) of teachers and 93% (93%) of students never used calculators in their classes. And, 93% (96%) of teachers and 96% (92%) of students never used computers in their math classes. This phenomenon was very unexpected because at the time of the TIMSS survey every elementary and secondary school had already over 50 computers and Korea was one of the top five computer manufacturing countries. Some teachers still have the belief that technology has no more positive effect on mathematics education than the traditional paper-and-pencil method. However, now, most math teachers agree that technology is an inevitable tool every student has to use in the huge stream of the current information society.

Fifth, the TIMSS report shows that many math teachers do not consider mathematical application in daily life and reasoning ability to support logical conclusions as important goals for students to achieve. Korea belonged to the lowest ranking countries in this category. This implies that teachers do not well understand the reason why problem solving is emphasized in the current curriculum. Although the percentage of teachers considering memorization of formulas and procedures to be important is lower than other countries, application of mathematics and reasoning should be emphasized more.

Sixth, ironically, the good results achieved in TIMSS made Korean mathematics educators reconsider their evaluation system. Except for a small portion of subjective items requiring short answers, most of the TIMSS questions were objective ones with five choices. In that sense critics have proposed that the objective items might have given a better chance for Korean students to get higher scores than European students because Korean students were familiar with objective tests. Alternative performance assessment items like projects and open-ended items might have produced worse results than the objective items that appropriated a large part in TIMSS.

THE 7TH NATIONAL CURRICULUM

The 7th curriculum, reflecting a movement that occurred within mathematics education throughout the world, particularly in the United States since the 1980s, focuses on the improvement of students' thinking abilities. This change stemmed from internal reflections on teaching and learning methods and contents of school mathematics that had been emphasized for a long time. School mathematics has been accused of being confined only to fragmentary concepts and skills, and not providing students with interesting or challenging situations (Lew, 1999). This change was the result of an awareness that such a restricted education was not enough in that the role of mathematics education is a factor in leading the developmental stages from an industrial society to an information society.

CHARACTERISTICS OF THE 7TH NATIONAL CURRICULUM AND MAIN FEATURES OF CHANGES IN PAST CURRICULA

The ultimate goal of the 7th mathematics curriculum is to cultivate students with a creative and autonomous mind by achieving the following three aims (Ministry of Education, 1997): (1) to understand basic mathematical concepts and principles through concrete experiences using various manipulative materials and the use of daily life phenomena related with mathematics; (2) to foster mathematical modeling abilities through the solving of various problems posed with and without mathematics; and (3) to keep a positive attitude about mathematics and mathematics learning by emphasizing a connection between mathematics and the real world.

The new curriculum, which emphasizes basic mathematical knowledge and application of mathematics focusing on practical mathematics, contrasts sharply with the 3rd curriculum, which emphasized "mathematical structure" focusing on theoretical mathematics. Clearly, such contrast could not occur simply after one revision. The 4th curriculum issued in 1980 revised the 3rd curriculum to strongly accept the philosophy of "New Math" which was designed for reflecting the pure mathematics that rapidly developed in the twentieth century. The 4th revision of the mathematics curriculum carefully selected contents of the 3rd curriculum and emphasized basic skills, problem solving and an integrated approach to school subjects, which was influenced from the "Back to Basics" movement and problem-solving movement of the United States (H. S. Park, 1991). For about 20 years after 1980, the basic direction of mathematics education slowly changed from a focus on theoretical aspects to a focus on practical aspects such as problem solving, application, and the use of technology.

The 5th curriculum, which was a partial revision of the 4th curriculum, emphasized problem solving, mathematical activities and attitudes of students. Elementary school mathematics textbooks of Grades 1-6 included units titled "various problems," emphasizing various problem-solving strategies like simplification, logical reasoning, making a figure, working backward, and so on. The 6th curriculum emphasized problem solving even more and introduced computer use into the mathematics. In the 7th mathematics curriculum, students are expected to be able to organize

real-world phenomena mathematically, to determine mathematical relations of concepts and principles by the process of abstraction based on
their own concrete operations, to promote mathematical reasoning abilities by way of solving various problems using mathematical knowledge
and skills they have already acquired, and finally to acquire a positive attitude toward mathematics (Woo, 1992).

Traditionally, the most serious problem in Korean mathematics education is that mathematics is considered merely as a tool, a subject for students to prepare for the college entrance examination in which
theoretical mathematics, focusing on mathematical knowledge, is emphasized more than practical mathematics, focusing on its utility (Woo, 1992).
Such emphasis on test preparation without internal motivation for learning has made it difficult for students to obtain a real understanding and
to develop reasonable and productive thinking abilities. It demands students to mechanically accept undigested content organized around topics
frequently appearing in the examination. As a result, it becomes almost
impossible for students to nurture an investigative attitude and other
desirable mental habits.

Korea has a long history of emphasizing theoretical mathematics in
selective examinations (see Figure 4.2). In A.D. 958 the *Koryeo* Kingdom
(A.D. 918-1392) introduced mathematics into its examination system to
select government officials (Needam, 1954, p. 139). Whoever wanted to
become a government technical official was required to have learned
many Chinese mathematical classics including *Chui-chang Suan Shu* (Nine
Chapters on Mathematical Art), for 7 to 9 years in the national schools.
This tradition of mathematics education had continued during the period
of the *Koryeo* Kingdom and *Chosun* Kingdom (A.D. 1392–1910).

INSTRUCTION AND EVALUATION

The 7th curriculum emphasizes various types of instruction to improve
efficiency and a sense of significance in students' mathematical learning
(Ministry of Education, 1997). It recommends to teachers that students

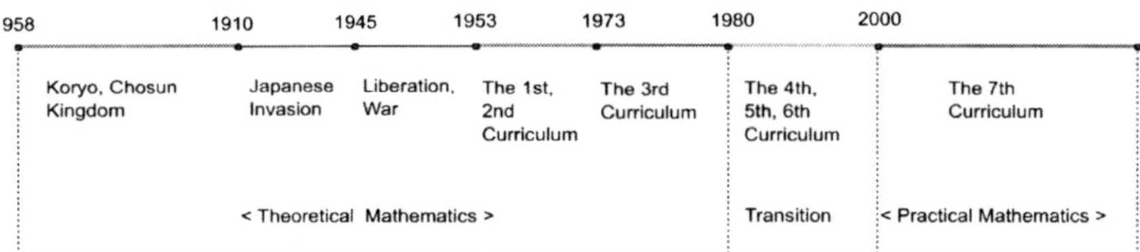

Figure 4.2.   History of mathematics curriculum of Korea.

should be able to experience the joy of discovery and maintain their interest in mathematics by pursuing the following instructional methods in their classrooms:

- to emphasize concrete operational activities in order to help students discover concepts and rules and solve problems embedded in such a discovery;
- to have students develop basic skills and problem-solving abilities in order to use mathematics in their everyday life;
- to present concepts and principles in the direction from the concrete to the abstract in order to activate self discovery and creative thinking;
- to induce students to recognize and formulate problems from situations both within and outside mathematics;
- to select appropriate questions and subsequently provide feedback in a constructive way in order to consider the stages of students' cognitive development and experiences;
- to use open-ended questions in order to stimulate students' creativity and divergent thinking;
- to value the application of mathematics in order to foster a positive attitude toward mathematics; and
- to help students understand the problem solving process and use basic problem-solving strategies in order to enhance students' problem-solving abilities.

The 7th curriculum asserts that mathematical power should be evaluated by realizing the following evaluation methods in classrooms (Ministry of Education, 1997):

- to emphasize processes more than products in order to foster students' thinking abilities;
- to focus on students' understanding of a problem and the problem-solving process as well as its results in order to evaluate students' problem-solving abilities;
- to focus on students' interests, curiosity and attitudes toward mathematics in order to evaluate students' mathematical aptitudes;
- to focus on student's abilities to think and solve problems in a flexible, diverse and creative fashion in order to evaluate mathematical learning;
- to use a variety of evaluation techniques such as extended response questions, observations, interviews as well as multiple-choice questions in order to evaluate students' mathematical learning.

THE KOREAN EDUCATIONAL SYSTEM

Korea has a 6-3-3-4 educational system comprised of 6 years of elementary, 3 years of middle school, 3 years of high school, and 4 years of college. According to educational statistics for the year 2004, there are about 589,000 students in each grade of high school (411,000 for general high schools and 172,000 for vocational high schools) and almost 99% of students graduate from high school schools (Korea Educational Development of Education, 2005). In the general high schools, students choose between a liberal arts track and a science track at the 11th grade. The ratio of students within these different tracks is about two to one respectively.

GENERAL STRUCTURE OF THE CURRICULUM

Table 4.1 shows the general structure of the 7th curriculum. The curriculum has two parts (Ministry of Education, 1997). One is the compulsory core curriculum used in the first 10 school years (the "people's common educational period"), which is divided into 20 sublevels. The compulsory core curriculum was introduced to teach the same content to all students. The second part consists of various elective courses on different topics and with different difficulty levels like math I, math II, calculus, real world mathematics, statistics and probability, and discrete mathematics. Students in the 11th and 12th grades may opt for those according to their interests, abilities and future direction. The elective curriculum has two kinds of selective subjects: General elective subjects designed to strengthen cultural literacy and connection to the real world (real world mathematics) and deep elective subjects that are centered around five subjects related to the college entrance examination (math I, math II, calculus, statistics and probability, and discrete math).

Generally, a student who fails a level transfer test by scoring below 60% must remain at that level and take the course again. Some exceptions will be made, by request of the student, the parent, or by way of the teacher's decision, such that failing students may transfer to the next level, in which case a special course will be administered by the school at the end of the semester, operated according to the minimum criteria laid down by the national curriculum.

A given level in the common core may be repeated only once. Students who fail a second time have to move to the next level. To avoid unnecessary competition and a sense of incongruity among students and parents, skipping levels is not permitted. Instead, for advanced students and low achievers, special enrichment and supplementary sections will be pro-

Table 4.1. Structure of Mathematics Curriculum

School	Elementary School												Middle School						High School			
Grades	1		2		3		4		5		6		7		8		9		10		11	12
Levels	1 a	1 b	2 a	2 b	3 a	3 b	4 a	4 b	5 a	5 b	6 a	6 b	7 a	7 b	8 a	8 b	9 a	9 b	10 a	10 b		
Curriculum Component	Compulsory Core																				Elective	
Subject	Mathematics																				(General Electives) Real World Math	
																					(Deep Electives) Math I, Math II, Calculus, Probability & Statistics, Discrete Math	

Table 4.2.   Time Allotment

School	Elementary*						Middle**				High***	
Grade	1	2	3	4	5	6	7	8	9	10	11	12
Curriculum	Compulsory Core Curriculum										Selective Curriculum	
Class hours for 1 year (34 weeks)	120	136	136	136	136	136	136	136	102	136	Liberal arts 102 Science 272	Liberal arts 102 Science 136
Class hours per week	3-4	4	4	4	4	4	4	4	3	4	Liberal Arts 3 Science 8	Liberal Arts 3 Science 4

*Class hour of elementary school is 40 minutes.
**Class hour of middle school is 45 minutes.
***Class hour of high school is 50 minutes.

Table 4.3.   Class Hours Classified by Strands

Strand	Class Hours	Ratio
Numbers and operations	448	34.2%
Geometry	290	22.1%
Measuring	158	12.1%
Probability and statistics	98	7.5%
Letters and expressions	197	15.0%
Patterns and functions	119	9.1%
Total	1,310	100%

Table 4.4.   Class Hours Classified by Strands and Grades

Strands	Grade 1	Grade 2	Grade 3	Grade 4	Grade 5	Grade 6	Grade 7	Grade 8	Grade 9	Grade 10
Numbers and operations	*74	63	80	63	59	32	34	9	13	21
	62	46	58	46	44	24	25	7	13	15
Geometry	13	22	25	35	32	26	34	48	30	25
	11	16	18	26	24	19	25	35	29	18
Measuring	14	24	15	17	20	23	18	8	11	8
	11	17	11	12	14	17	13	6	11	6
Probability and statistics	6	7	8	10	9	18	13	14	7	6
	5	5	6	7	6	13	10	10	7	4
Letters and expressions	9	17	6	8	12	14	22	39	27	43
	8	12	4	6	9	10	16	28	27	31
Patterns and functions	4	3	2	3	4	23	15	18	14	33
	3	2	1	2	3	17	11	13	14	24
Sum	120	136	136	136	136	136	136	136	102	136
	100	100	100	100	100	100	100	100	100	100

*In each cell above, the top number is the number of class hours, the lower number is the percentage of total class hours.

vided during the course at the teacher's discretion. This system is a conversion from the current policy, which allows automatic grade promotion regardless of a student's achievement scores, to a policy which controls level transfer according to achievement criteria set up in advance.

Table 4.2 shows that time allotted to mathematics in Grades 1-10 are on average 131 class hours per year. It is the second largest time allot-

**Table 4.5.   Important Concepts of the
Six Strands and Grades in Which They are Taught**

Strand	Important Concepts	Elementary						Middle			HS
		1	*2*	*3*	*4*	*5*	*6*	*7*	*8*	*9*	*10*
Numbers & operations	Natural number	X	X	X	X	X		X			
	Fraction and decimal fraction			X	X	X	X				
	Integer and rational							X	X		
	Real and complex									X	X
Geometry	Spatial sense	X	X	X	X	X	X	X			
	Properties of figures		X	X	X	X	X	X	X		
	Deductive proof								X	X	
	Coordinate geometry										X
Measuring	Measuring with units	X	X	X	X	X	X	X			
	Quantity sense and estimation	X	X	X	X		X		X		
	Trigonometric ratio									X	
	Linear programming										
Probability & statistics	Diagram and graphs	X	X	X	X	X	X				
	Probability						X		X		
	Statistics					X		X		X	X
Letters & expressions	Making expression	X						X	X	X	X
	Equation & inequality		X					X	X	X	X
	Factorization									X	X
	Problem-solving strategy	X	X	X	X	X	X				
Patterns & functions	Pattern finding	X	X	X	X	X					
	Ratio and proportion						X	X			
	Elementary function							X	X	X	X
	Trigonometric function										X

Table 4.6.   Class Hours Classified by Strands and Grades (Numbers and Operations)

| Important Concepts | Elementary | | | | | | Junior High | | | SH | |
	1	2	3	4	5	6	7	8	9	10	Total
Natural	*74	63	70	27	17		18				269
numbers	62	46	51	20	13		13				20.5
Fractions &			10	36	42	32					120
decimal			7	26	31	24					9.2
fractions											
Integers &							16	9			25
rational							12	7			1.9
numbers											
Real &									13	21	34
complex									13	15	2.6
numbers											
Sum	74	63	80	63	59	32	34	9	13	21	448
	62	46	58	46	44	24	25	7	13	15	34.2

*In each cell above, the top number is the number of class hours, the lower number is the percentage of total class hours.
Class hours per year: 1,310 (1st grade, 102; 9th grade, 120; others, 136

Table 4.7.   Class Hours and Percentages of Class Time, Classified by Strands and Grades (Geometry)

| Important Concepts | Elementary | | | | | | Junior High | | | SH | |
	1	2	3	4	5	6	7	8	9	10	Total
Spatial	*13	16	11	2	4	8	13				67
sense	11	12	8	1	3	6	10				5.1
Properties		6	14	33	28	18	21	24			144
of figures		4	10	24	19	13	15	18			11.0
Deductive								24	30		54
proof								17	29		4.1
Coordinate										25	25
geometry										18	1.9
Sum	13	22	25	35	32	26	34	48	30	25	290
	11	16	18	26	23	19	25	35	29	18	22.1

*In each cell above, the top number is the number of class hours, the lower number is the percentage of total class hours.
Class hours per year: 1,310 (1st grade, 102; 9th grade, 120; others, 136).

Table 4.8.   Class Hours and Percentages of Class Time, Classified by Strands and Grades (Measuring)

| Important Concepts | Elementary | | | | | | Junior High | | | SH | |
	1	2	3	4	5	6	7	8	9	10	Total
Measuring with units	*6	22	13	10	20	17	18				106
	4	16	10	7	14	13	13				8.1
Quantity sense and estimation	8	2	2	7		6		8			33
	7	1	1	5		4		6			2.5
Trigono-metric ratios									11		11
									11		0.8
Linear pro-gramming										8	8
										6	0.6
Sum	14	24	15	17	20	23	18	8	11	8	158
	11	17	11	12	14	17	13	6	11	6	12.1

*In each cell above, the top number is the number of class hours, the lower number is the percentage of total class hours.
Class hours per year: 1,310 (1st grade, 102; 9th grade, 120; others, 136).

Table 4.9.   Class Hours and Percentages of Class Time, Classified by Strands and Grades (Probability and Statistics)

| Important Concepts | Elementary | | | | | | Junior High | | | SH | |
	1	2	3	4	5	6	7	8	9	10	Total
Diagram and graphs	*6	7	8	10	6	8					45
	5	5	6	7	4	6					3.4
Probability						10		14			24
						7		10			1.8
Statistics					3		13		7	6	29
					2		10		7	4	2.2
Sum	6	7	8	10	9	18	13	14	7	6	98
	5	5	6	7	6	13	10	10	7	4	7.5

*In each cell above, the top number is the number of class hours, the lower number is the percentage of total class hours.
Class hours per year: 1,310 (1st grade, 102; 9th grade, 120; others, 136).

Table 4.10.   Class Hours and Percentages of Class Time, Classified by Strands and Grades (Letters and Expression)

| Important Concepts | Elementary | | | | | | Junior High | | | SH | |
	1	2	3	4	5	6	7	8	9	10	Total
Making	*2						10	14	7	17	50
expression	2						7	10	7	13	3.8
Equation &		7					12	25	13	21	78
inequality		5					9	18	13	15	6.0
Factoriza-									7	5	12
tion									7	4	0.9
P.S. strategy	7	10	6	8	1	14					57
	6	7	4	6	9	10					4.4
Sum	9	17	6	8	12	14	22	39	27	43	197
	8	12	4	6	9	10	16	28	27	31	15.0

*In each cell above, the top number is the number of class hours, the lower number is the percentage of total class hours.
Class hours per year: 1,310 (1st grade, 102; 9th grade, 120; others, 136).

Table 4.11.   Class Hours and Percentages of Class Time, Classified by Strands and Grades (Patterns and Functions)

| Important Concepts | Elementary | | | | | | Junior High | | | SH | |
	1	2	3	4	5	6		8	9	10	Total
Pattern	*4	3	2	3	4						16
finding	3	2	1	2	3						1.2
Ratio &						23	6				29
proportion						17	4				2.2
Elemen-							9	18	14	15	56
tary func-							7	13	14	11	4.3
tion											
Trigono-										18	18
metric										13	1.4
function											
Sum	4	3	2	3	4	23	15	18	14	33	119
	3	2	1	2	3	17	11	13	14	24	9.1

*In each cell above, the top number is the number of class hours, the lower number is the percentage of total class hours.
Class hours per year: 1,310 (1st grade, 102; 9th grade, 120; others, 136).

**Table 4.12. Important Concepts of the
Selective Mathematics Curriculum of Grades 11–12**

Electives	*Important Concepts*
Real world mathematics	Calculator and computer; economy and life; statistics in ordinary life; real-life problem solving
Mathematics I	Exponent and logarithm; matrix; sequence; permutation & combination; probability statistics
Mathematics II	Equations & inequalities; limit and continuity of a function; differentiation; integration; quadratic curves; space figures; coordinates in space; vectors; limits of a function; differentiation
Differentiation & integration	Trigonometric functions; limits of a function; differentiation; integration
Probability & statistics	Descriptive statistics; probability
Discrete mathematics	Selections and arrangements; graph theory; algorithms; decision making and optimization

Table 4.13. Korea Scholastic Ability Test

Type	*Subjects*	*Ratio of Subjects*	*Number of Items*	*Running Time*	*Type of Items*
A	Math I, Math II, Select one subject among calculus, statistics & probability, discrete mathematics	Math I 40% Math II 40% Electives 20%	30 items	100 minutes	Multiple Choice:70%, Short Answer:30%
B	Math I	Math I			

ment, behind Korean language, of which the time allotted is on average 188 hours per year. In Grades 11 and 12, the time allotment depends on which track for the college entrance examination the student is in: science or liberal arts.

In the case of the science track, students learn 8 hours of math I in the first semester of 11th grade and 8 hours of math II in the second semester of 11th grade. In the first semester of 12th grade, one of three subjects, calculus, probability and statistics, and discrete math, is studied for 4 class hours weekly and in the second semester of 12th grade, another of these subjects is studied for 4 class hours weekly. In the case of the liberal arts track, in Grade 11, math I is learned for two semesters for 3 hours per week and in Grade 12, probability and statistics is learned for two semesters for 3 class hours per week.

Table 4.14.   Revision Cycle of National Curriculum

Curriculum	Level (Date)	Laws/Decrees
Temporary curriculum	Elementary (Sep 1946) High (Sep 1947)	Notification by U.S. Occupational Forces
1st	Elementary, High (Aug 1955)	Decree by the Ministry of Education No. 45
2nd	Elementary, Middle, High (Feb 1963)	Decree by the Ministry of Education No. 119, 120, 120
3rd	Elementary (Feb 1973) Middle (Aug 1973) High (Dec 1974)	Decree by the Ministry of Education No. 310, 325, 350
4th	Elementary, Middle, High (Dec 1981)	Decree by the Ministry of Education 442
5th	Elementary (Jun 1987) Middle (Mar 1987) High (Mar 1988)	Decree by the Ministry of Education 87-9, 87-7, 88-7
6th	Elementary (Sep 1992) Middle (Jun 1992) High (Oct 1992)	Notification by the Ministry of Education, No. 1992-16, 1992-11, 1992-19
7th	Elementary, Middle, High (Dec 1997)	Notification by the Ministry of Education No. 1997-15
New	(A plan) Grade 1-10 (Dec 2005) Grade 11-12 (Dec 2006)	

THE MATHEMATICS CURRICULUM OF GRADES 1-10

The "Compulsory Period" mathematics curriculum consists of the following six strands: numbers and operations, geometry, measuring, probability and statistics, letters and expressions, and patterns and functions. Unlike the United States, Korea does not distinguish between the content strand and process strand. The only exception is problem solving, which is included in the letters and expressions strand. Table 4.3 shows class hours classified by the six strands along with their ratios. According to Table 4.3, numbers and operations (34.2%) and geometry (22.1%) have greater emphasis than probability and Statistics (7.5%) and patterns and functions (9.1%).

Table 4.4 displays the class hours and its percentage that is being taught in each strand and each grade level. According to Table 4.4, the National Curriculum of Korea teaches these six subject strands in all grade levels 1–10. From 1st grade to 5th grade, numbers and operations and geometry combined, the percentage appears to be quite high as demonstrated in the following numbers: 73%, 62%, 76%, 72%, 68%. On the contrary, Probability and Statistics and Patterns and Functions are not taught as much. Although all areas are taught evenly in the sixth and seventh grades, Geometry and Letters and Expressions account for 63% and 56% respectively in the 8th and 9th grades, while Letters and Expressions and Patterns and Functions are highly emphasized in the 10th grade.

Table 4.5 shows 23 important concepts in the six strands and the years in which they are taught. A notable point is that geometry is taught as a separate subject in eighth and ninth grades. Although proofs are also covered in part through spatial geometry at the high school level, the geometry that is covered in high school is mostly coordinate and vector. Owing to the fact that the achievement level demonstrated by students is so low, whether or not geometry should be taught in middle school is a quite controversial issue in Korea. Many scholars argue that it should be moved to high school, but a conclusion has yet to be reached due to the long-standing custom of teaching it in middle school. The problem-solving strategies in the Letters and Expressions strand are similar to the problem-solving strategy strands presented in U.S. standards. However, it is not presented as an independent strand, but dealt with as an important concept in the content strand of letters and expressions.

Table 4.6 presents the class hours and time percentages that are taught in each grade, categorized by the four important concepts that are covered in the numbers and operations strand. In the first, second and third grades the focus is set on natural numbers; in the fourth, fifth, and sixth grades students learn concepts accompanying fractions and decimals; in the eighth grade constant and rational numbers are looked at while in the ninth and 10th grades real and complex numbers are dealt with. Such sequence reflects a unique characteristic of the Korean national curriculum that deals with main big ideas in sequence.

Table 4.7 displays the class hours and the percentage of time in which they are taught in each grade, categorized by the four important concepts dealt with in the geometry strand. Ideas emphasized are: in the first and second grades, spatial sense, in the third through eighth grades, the properties of figures, in the eighth and ninth grades, deductive proofs, and in 10th grade coordinate geometry. The spatial sense that is promoted in the seventh grade has its purpose in fortifying intuitive activities prior to the instruction of logical proof in the eighth grade. However, despite such preparatory activities, the treatment of proofs that is presented in seventh and eighth grades poses a great cognitive challenge to students. Rather, the coordinate geometry that is presented in the 10th grade is, in actuality, perceived to be easier. This issue of the location of the geometry strand has always been a point of discussion during revision of the national curriculum.

Table 4.8 displays that class hours and the percentage of time that are used to instruct at each grade level, categorized by the 4 important concepts dealt with in the measurement strand. In this strand various measuring activities and conversion of measurement units including weight, capacity, time, angle, volume, area, length, etc. are utilized.

Although measurement-related topics are taught all throughout the period from first to seventh grade, they are concentrated in 2nd, 3rd, 5th, 6th and 7th grades. Mass and volume, and estimation are taught generally, except in fifth and seventh grades, however, the percentage of instruction time within the overall mathematical context is quite low. It can be concluded that mass and volume, and estimation are relatively neglected when compared to U.S. standards. In ninth grade, trigonometric ratios are taught in terms of geometry and not in terms of functions and in 10th grade basic linear programming is covered.

Table 4.9 presents the class hours and their percentages that are covered in each grade level categorized by the three big ideas dealt with in the probability and statistics strand. Although tables and graphs are continuously dealt with from Grades 1–6, it seems that they are not particularly emphasized. Probability and statistics is instructed alternately starting from Grade 5.

Table 4.10 displays the class hours and the percentages of time that are devoted to covering important concepts dealt with in the letter and expression strand. In the case of equations, that are instructed in grades one and two, although the topic of equations itself may not appear as much, it forms an imperative foundation for dealing with the numbers and operation strand and is also closely connected with the problem-solving strategies that are continuously instructed from first grade through to sixth grade. Manipulation of letters and expressions is specifically instructed at the middle school level. From Grades 7–10, manipulation of equations and inequalities is systematically taught. Factorization of equations of second degree is taught at the ninth grade level, and in the 10th grade level factorization of equations of higher degrees is taught.

Table 4.11 displays the class hours and its percentage that are used to deal with the important concepts that are instructed in the patterns and functions strand.

From Grades 1–5, pattern finding is instructed, in Grades 6–7 percentage and proportional expression is taught, while in the seventh grade functions are introduced based on the concepts of constant and inverse proportion. Linear functions are touched on in the eighth grade, functions of the second degree are dealt with in the ninth grade, while rational functions and trigonometric functions are taught in the 10th grade.

CONTENTS 11TH AND 12TH GRADES

This "selective" curriculum is applied to students in Grades 11–12. In these two grades various math subjects are available such as practical mathematics, mathematics I, mathematics II, calculus, probability and statistics, and discrete mathematics. Table 4.12 shows important concepts of the elective mathematics curriculum of Grades 11–12.

Practical mathematics is an optional course offered to students who want to learn mathematics for daily life without having to complete the 10th grade level. This subject enables students to apply the basic concepts and rules of mathematics, to ultimately consider various types of problem solving in real-life situations.

Mathematics I is the first course to be offered to students who wish to study advanced mathematics after completing level 10 of mathematics in the compulsory period. Through this course, students are able to understand basic mathematical concepts, principles, and laws, and develop mathematical thinking ability, logical reasoning ability, and reasonable and creative problem-solving ability.

Mathematics II is a course to be offered to students who want to study more advanced mathematics after mathematics I. Through this course students can attain deeper mathematical knowledge and better develop their mathematical thinking ability, logical reasoning ability, and then develop abilities and attitudes to solve problems appropriately. This course is suitable for students who wish to study natural sciences or technological sciences at the college level.

Differentiation and integration is a course designed for students who want to study advanced differentiation and integration of various functions after having completed mathematics II. In this course students will be able to gain advanced knowledge in differentiation and integration. They will develop their mathematical thinking, logical reasoning, and problem-solving ability. This course is appropriate for students who want to study natural sciences or technology at the college level.

Probability and statistics is an optional course offered to students who wish to study applied probability and statistics without having to complete level ten mathematics. This subject enables students to improve their data processing ability and their inferential ability necessary for the information age. It will enable them to understand the statistical phenomena in society and nature and, hence, to improve their analytical ability. It is suitable for students who need to use probability and statistics in real-life situations through experimental and operational activities.

Discrete mathematics is offered to students regardless of whether they have completed level 10 mathematics or not. In discrete mathematics, the use of basic mathematical concepts, principles, and laws will develop a student's abilities and aptitude to analyze mathematically, to think logically and to solve reasonably finite or discontinuous discrete problem situations. This is a course needed for students who want to have experience in discrete mathematical knowledge.

THE ROLE OF INSTRUCTIONAL TECHNOLOGY IN KOREA'S NATIONAL CURRICULUM

The 7th mathematics curriculum recommends in its "guidelines for instruction" that "computers and calculators should be actively utilized to improve understanding of concepts and problem solving or thinking abilities in classrooms." This is the only sentence that shows the role technology plays in the curriculum. It is according to this guideline that all textbooks must contain at least some mathematical activities that involve technology. In Korean textbooks, technology is being used even within limited bounds.

- A drawing tool for function graphs (TI graphing calculator) and geometric figures under some conditions (GSP) without evoking students' thinking (Figure 4.3)
- A calculation tool for complex and difficult calculation (Scientific Calculator) (Figure 4.4).
- A calculation tool for understanding the given theorem or formula more clearly (Excel) (Figure 4.5).

However, a problem lies in teachers' preconceptions about technology. Technology has had a dual position in that many mathematics teachers consider technology to be very helpful while many others also consider it to be very harmful. The computer is helpful as well as harmful depending on how it is used. Therefore the methods appropriated for using the computer should be changed.

Various activities such as the ones set forth below should be developed:

- Activities to find mathematical invariance through investigations of numerical relations.
- Activities to connect multiple representation functions of graphs, expressions, diagrams and real situations for understanding concepts.
- Activities to get key ideas in special problems in which solution is extremely difficult or impossible without computers.

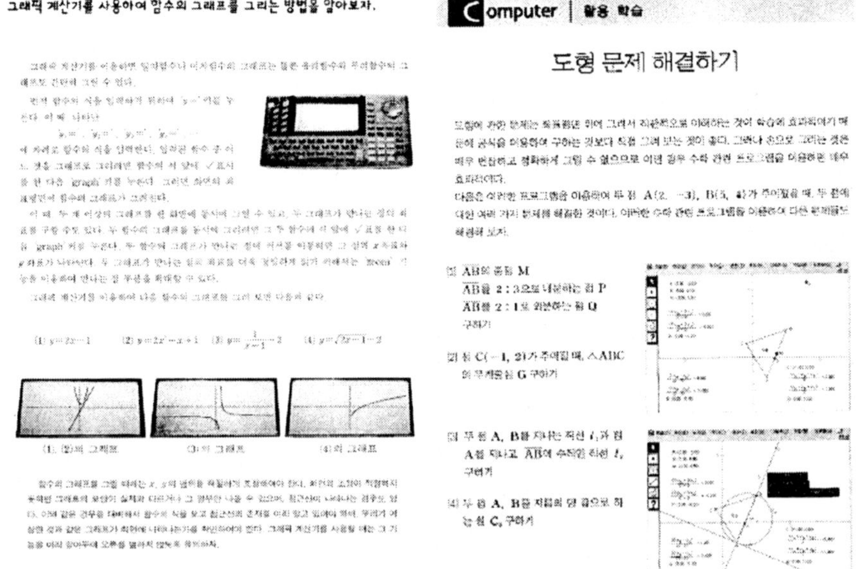

Sources: Left: Choi et al. (2002, p. 151). Right: Shin and Choi (2002, p. 79).

Figure 4.3.   Use of TI graphing calculator and Geometer's Sketchpad in Korean textbooks.

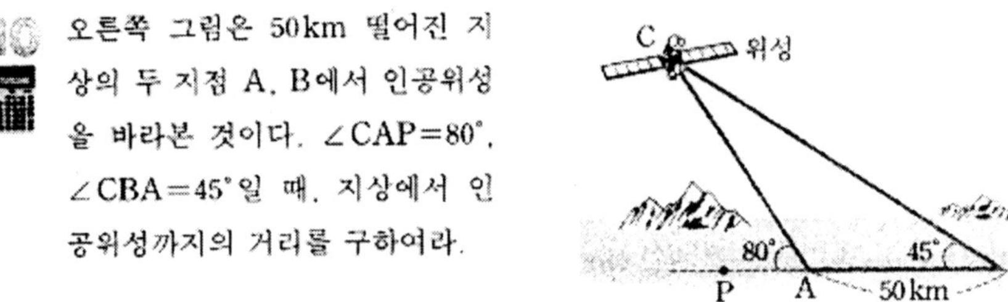

Source: Choi et al. (2002, p. 204).

Figure 4.4.   Use of a scientific calculator in a Korean textbook.

The 7th curriculum was designed for teachers to be able to teach without a computer. All of the content can be learned without a computer, although teachers and students are permitted to use them in their classes.

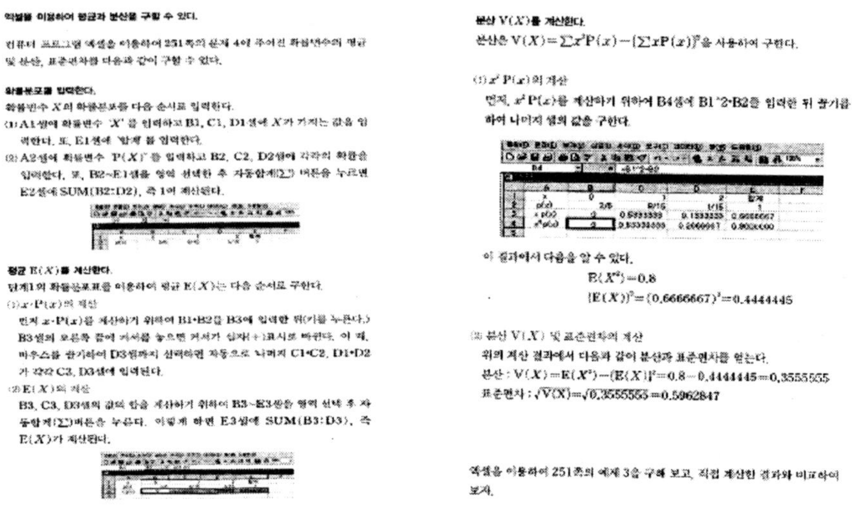

Source: S.H. Yim, Gee, Kim, and Lee (2002, pp. 225–253).

Figure 4.5.   Use of Excel in a Korean textbook.

The use of computers depends on the interests and abilities of individual teachers, but many teachers are not rising to this challenge because they do not necessarily feel the need to use computers. Therefore, computers should be necessitated in the mathematics curriculum rather than just appear on some pages of the textbooks in the sense of the following three aspects: Computer should be mandatory; many mathematical contents can be learned more meaningfully than in a paper-and-pencil environment; and various mathematical activities with computers should be evaluated in the formal examination. In Korea, it is difficult for teachers to use computers in their classrooms because computers are not mandatory and they do not know the proper way to use computers in their examinations. Even though teachers use computers in their classrooms, students acknowledge that activities with computers are not evaluated.

THE KOREA SCHOLASTIC ABILITY TEST
AND THE NATIONAL CURRICULUM

The Korea Scholastic Ability Test is a state-supervised test for high school graduates or expectant graduates who want to enter college. Its purpose is to measure a student's scholastic ability for college education. Most universities use the score of this test in the process to select their new students. Students can select the type and subject of the test, based on the regulations of the university to which they want to apply. Students select one out of two types of the test (type A and type B) in mathematics (see Table 4.13). In the case of type A, students can select one subject among three electives of calculus, statistics and probability, and discrete mathematics. Test items will be focused on 11th and 12th grade subjects. Since the contents of the 11th and 12th grade subjects are based on the previous subjects learned until 10th grade, it is feasible that test items would be indirectly related to those previous subjects.

THE REVISION PROCESS OF KOREA'S MATHEMATICS CURRICULUM REVISION CYCLE AND BACKGROUND

Table 4.14 shows the different periods in which revisions were made to Korea's national curriculum. According to this table, Korea's revisions were made every 7 to 10 years until the 4th curriculum and after that until the 7th curriculum every 5 to 7 years on average. The currently pending revision of the 7th national curriculum will be done approximately 8 to 9 years after the last revision, since revision for grade levels under 10th grade was scheduled for December 2005, while revision for grade levels 11 through 12 was scheduled for December 2006.

COMPOSITION OF CURRICULUM REVISION TEAMS

Once the Ministry of Education selects a group to revise the curriculum, this group first embarks on research into previous studies. The group that revised the 6th curriculum was comprised of a group of seven university professors from different schools (Kang et al., 1997). This committee carried out basic required research using the budget allocated to them. The group commissioned by the MOE toward revising the 7th curriculum is The Korea Institute of Curriculum and Evaluation (KICE). KICE is a national research center that was established by the national government and carries out curriculum-related research and oversees national level tests and exams. Two university professors were added to a group of seven researchers to form the group of nine to conduct this curriculum research.

FACTORS INFLUENCING REVISION OF THE MATHEMATICS CURRICULUM

It is observed that the factors that influence the curriculum are not so different from those of other nations. The following few pages will mainly discuss the transfer process from the 6th to the 7th curriculum and the recent and current process of modifying and transforming the 7th curriculum. The aforementioned groups carry out basic required research and conduct surveys pertaining to the following items in order to change the framework or content for the new curriculum.

Analysis of Problems in Applying the Existing Curriculum

In the situations that the existing curriculum fails to lead to the mathematical education of students, the system is modified. For example, the level-based instruction that is embedded in the 7th curriculum was heavily criticized because of the difficulties with its execution (KICE 2005a, 2005b), thus this level-based approach will not be continued for the new curriculum. The system goes through fervent analysis through discussions among teachers, practitioners, researchers, the MOE, and human resource development, professional associations, and so on.

Students' Level of Understanding

There is a NAEP-style examination that measures the level of student achievement (Cho, Lee, & Lee, 2004). The results of this examination provide information on how to perceive content as difficult or easy.

Logical Systemization of Subject Content

When teachers submit opinions on what worked and what did not work, the groups of researchers deliberate such opinions to analyze what problems there are with such teaching methods. The group of researchers goes through a process of negotiation with their collaborators to discuss desirable teaching methods.

International Comparisons

After conducting research on the main contents of the curricula of the United States, United Kingdom, Japan, China, France, Germany, and others, the items that are judged to be desirable are scrutinized for potential introduction into the Korean system (Yim, Lee, Lee, Park, & Jeong, 2004).

National and Societal Needs

Prior to revising the curriculum, what is perceived to be the necessary mathematical knowledge or mathematical thinking within the nation and society is investigated (S. K. Park et al., 2004).

Demands from Academia

Prior to revising the curriculum, a seminar is held in order to establish directions for revision of the curriculum with representatives from the academy (for example, KICE, 2005a, 2005b; KICE & Korea National University of Education, 2005).

Appropriateness of the Volume of Study

Whether or not the volume of material required to be learned by students is excessive or not is studied (J. H. Yim et al., 2004). In the case that it is deemed inappropriately lengthy, decisions are made on what is discarded and adjusted.

CURRICULUM REVISION PROCESS

Developing a mathematics curriculum at the national level is most certainly not an easy job. The process occurs on two levels comprised of general and detailed development and each level constitutes an individual project that spans approximately 1 year. The final version goes through a process of deliberation and revision by many education experts, teachers, parents and students, until it is consequently confirmed. The following are the stages involved in developing general remarks on the curriculum (Ministry of Education, 1999):

1. Establishment of a basic curriculum revision plan
2. Research on improvement of curricular system and structure
3. Fundamental research on curriculum revision

    a. Analysis, assessment and evaluation of current curriculum
    b. International comparison of curricula research
    c. Survey on opinions and demands by students, teachers, parents
    d. Survey on demands on the national and societal level and on prospects of public school education
    e. Survey on status of curriculum operation

4. Development of drafted revision of general remarks of curriculum

    a. Draft production and evaluation
    b. First round committee on revision and evaluation
    c. Second round committee on revision and evaluation
    d. Draft of revised general remarks

5. Forming of Curriculum Deliberation Committee

   a. Forming of Curriculum Deliberation Committee
   b. First round deliberation
   c. Revision according to evaluation
   d. Second round deliberation

6. Revised General Remarks

   a. Collection of opinions on drafted revision of general remarks

      i. Deliberation of elementary, middle school, high school sites
      ii. City and province boards of education, related academic associations and organizations, research organization, governmental departments

   b. Public hearing on drafted revision of general remarks
   c. Revision of general remarks

7. Confirmation of general remarks of curriculum

   a. Confirmation on fundamental direction of revision, revision points, main revised items of curriculum
   b. Confirmation of organization and standards in terms of time allotment
   c. Publishing and issuance of the revised general remarks of the curriculum and publicity information

Once the general remarks on the curriculum are confirmed, the details of the mathematical curriculum go through a process of revision. Similarly with the general remarks, the process of revising the particulars also involves the collection of opinions from experts, teachers, students, and parents. In order for the curriculum to be confirmed, a public hearing is held so that the opinions of many more people are voiced and can be taken into account.

1. Research on drafted revision of the mathematical curriculum

   a. Establishment of fundamental direction for mathematical curriculum development
   b. Analysis and evaluation of mathematical curriculum
   c. International comparative analysis of mathematical curricula
   d. Survey on the demands and opinions on applied practice of mathematics curriculum of each subject area by students, practitioners, teachers, parents, and related experts
   e. Seminar and conference

       i.   Status deliberation conference

      ii.   Conference on adjustment of development of general outline and detailed particulars

  f.   Production of draft of revised detailed contents of the mathematical curriculum

  g.   Revisions applied

2.   Deliberation of mathematical curriculum in practice and public hearing

  a.   Collection of opinions on drafted revision

       i.   Deliberation of practice in primary and secondary schools

      ii.   City and province boards of education-related association, different governmental departments

  b.   Public hearing on drafted revision

  c.   Revisions applied

3.   Confirmation of revised mathematical curriculum

4.   Follow-up support on revision of mathematical curriculum

  a.   Development of new mathematics books

  b.   Professional training in preparation of implementation of new curriculum

  c.   Publication of instructional manual on new curriculum

  d.   Publication and issuance of research data and publicity information

  e.   Production of guidelines to organization and implementation of curriculum by city and province

REVISION STATUS OF THE NEW CURRICULUM

One of the fundamental directions of the curriculum revision is that revisions need be pursued so that level-based instructions are carried out in an efficient way and that the curriculum can be executed appropriately to the present reality of Korea. Correspondence with student levels in the 7th curriculum was carried out in two stages. The first was by dismissing students who were not on par with a certain grade level determined by an examination at the end of each level of curriculum, and the second was by achievement level based instruction with students of similar achievement. Among the two, the first stage was deemed to be unrealistic and immediately aroused dissension from members of the educational community and therefore ultimately became obsolete. The decision to abolish the age-level concept of curriculum in the current revision process is judged to be a very timely and appropriate one. The second measure, achievement-level based classes, was not welcomed by the educational community either. This curriculum revision tries to find the method to strengthen the level-based instruction with minimizing various problems which have been raised from the educational sites. However there are no determined plans that respond to this issue, not at least in the proposed draft. It seems there is a need to make detailed notion of how revisions will be carried in order to accommodate and make level-based instruction possible.

CONCLUSION

As shown through the TIMSS study results, Korean students' mathematical scores are world class. However, the same TIMSS report also reflects that students dislike mathematics and are fearful of applying mathematics. It is a serious problem to distance oneself from mathematics or to consider mathematics inappropriate in the twenty-first century when mathematics is only increasingly more important for everyone than in the past. If the number of people working in mathematics related fields is insufficient, a serious labor problem will occur in a high-tech society. The need for mathematics is ubiquitous in our modern society, for there is no one who does not use mathematics, from a cashier to a rocket scientist, members of our society are not shameful about their mathematical illiteracy.

Such a society that exhibits such hypocrisy cannot be expected to develop further. Measures need be made so that more students can concentrate on math, and leave school with a sense of confidence about the subject. With rare exception, all students should be able to attain the mathematical confidence and competence required for private and public activities, acquire the mathematical skills necessary for their careers, and also recognize the social and cultural importance of mathematics. Until recently, school math has been perceived as a mere tool in preparing for the college entrance examination, and as a result the number of students who believe there is no value in further studying mathematics has significantly increased.

Many people think that the mathematics they learned in elementary school suffices for surviving in society, and think that only those who are professionally associated with math should learn the necessary mathematical skills. Such practices or perceptions can no longer prevail. The need for mathematics has been increasing in daily life, in the overall society, and at the work place. Mathematics increasingly plays an important role for those who wish to succeed in their professional fields. Therefore the field of mathematical education should guarantee that students leave school with the confidence and competence that responds to the mathematical demands that are embedded in their future lives. Change is needed, since school mathematics has been considered only as a tool to raise results in the college entrance examination. Korea has a long history of unsound mathematics education. It is very difficult to redress this phenomenon. Many Korean mathematics teachers and educators are making attempts to remodel their math classes despite such hardships.

This paper has summarized the 7th national mathematics curriculum made in Korea after release of the TIMSS report. The main focus is on the development of mathematical thinking abilities. This is related to the national strategy to construct a highly developed country in the near future: Mathematics provides tools for propelling the development of science and for solving quantitative and qualitative problems faced by people in their lives. To use these tools properly and thoroughly, fragmentary knowledge and skills are insufficient. Korean mathematics educators have taken upon themselves the specific educational objective of providing students with proper mathematical thinking experiences in order to synthesize their knowledge and skills

REFERENCES

Beaton, A. E., Mullis, I. V. S., Martin, M. O., Gonzalez, E. J., Kelly, D. L., & Smith, T. A. (1996). *Mathematics achievement in the middle school years: IEA's third international mathematics and science study* (TIMSS). Boston: Boston College.

Cho, Y. M., Lee, D. H., & Lee, B. J. (2004). *National assessment of educational achievement in 2003* (in Korean) (Research Report RRE 2004-1-4). Seoul: Korea Institute of Curriculum and Evaluation.

Choi, B. D., Kang, O. G., Hwang, S. K., Lee, J. D., Kim, Y. O., Jeon, M. G., et al. (2002). *Mathematics 10-b*. Seoul: Joongang JinHung.

Educational Testing Service. (1989). *A world of differences: An international assessment of mathematics and science*. Princeton, NJ: Author.

Im, H., & Kim, J. G. (1995). *The third international mathematics and science study (TIMSS): National report* (in Korean). Seoul, Korea: National Board of Educational Evaluation.

Kang, O. K., Kim, W. K., Park, K., Park, Y. B., Paik, S. Y., Shin, H. S., et al. (1997). *A draft of the 7th mathematics curriculum for elementary school, middle school and high school* (in Korean). Seoul: Sung Kyun Kwan University 7th Mathematics Curriculum Developmental Committee.

Kim, Y. C., Yang, S. S., Kim, Y. H., & Lee, J. H. (2001). *Policy development for private tutoring expenditure reduction* (in Korean) (Research Report CR 2001-15). Seoul: Korea Educational Development Institute.

Korea Institute for Curriculum and Evaluation. (2005a). Seminar for investigating revision direction for the national mathematics curriculum (in Korean). Research Material ORM 2005-26. Seoul, Korea: Author.

Korea Institute for Curriculum and Evaluation. (2005b, October 13). Material for the public hearing on the revision of mathematics curriculum (in Korean). Seoul, Korea: Author.

Korea Institute for Curriculum and Evaluation & Korea National University of Education. (2005, June 10). *Issues and developmental strategy for curriculum revision* (in Korean). Academic Joint Seminar of KICE and KNUE at 2005 Exhibition for Educational and Human Resources. Cheong-Won, Korea; Choong-Buk, Korea: KICE & KNUE.

Korea Educational Development of Education. (2005). Educational Statistics System, Korea National Center for Education Statistics and Information (in Korean). Retrieved June 18, 2008, from http://std.kedi.re.kr/jcgi-bin/publ/publ_yrbk_frme.jsp?menuid=3.

Lee, S. S. (2002). Private education expenditure for children and economic well-being of household (in Korean). *Journal of the Korean Home Economics Association, 40*(7), 211-228.

Lew, H. C. (1999). New goals and directions for mathematics education in Korea. In C. Hoyles, C. Morgan, & G. Woodhouse (Eds.), *Rethinking the mathematics curriculum* (pp. 218–217). London: Palmer Press.

Ministry of Education. (1997). *The 7th mathematics curriculum* (MOE Notification No. 1997-15). Seoul, Korea: Author.

Ministry of Education. (1999). *A commentary of the 7th mathematics curriculum*. Seoul, Korea: Author.

Mullis, I. V. S., Martin, M. O., Beaton, A. E., Gonzalez. E. J., Kelly, D. L., & Smith, T. A. (1997). *Mathematics achievement in the primary school years: IEA's Third International Mathematics and Science Study*. Chestnut Hill, MA: TIMSS International Center, Boston College.

Needam, J. (1954). *Science and civilization in China, Vol. 1*. Cambridge, England: Cambridge University Press.

Park, H. S. (1991). Historical development of mathematics education in Korea. In J. H. Woo (Ed.), *Proceedings of the Korea/US Seminar on Comparative Analysis of Mathematics Education in Korea and the United States* (pp. 1-25). Seoul: Korea Society of Educational Studies in Mathematics.

Park, K. (2004, July 6). Factors contributing to Korean students' high achievement in mathematics. In H. C. Lew (Ed.), *The report on mathematics education in Korea* (pp. 85–92). The Korean Presentation at ICMI-10, Copenhagen, Denmark, Korea Sub-commission of ICMI.

Park, S. K., Heo, K. C., Lee, K. W., Kim, P. K., Lee, M. S., & Jeong, Y. K. (2004). *Survey on national and societal demands for revision of mathematics curriculum* (Research report CRC 2004-4-2). Seoul: Korea Institute of Curriculum and Evaluation.

Shin, H. S., & Choi, Y. J. (2002). *Mathematics 10-a*. Seoul, Korea: CheonJae.

Yim, S. H., Gee, H. S., Kim, D. S., & Lee, B. S. (2002). *Mathematics I*. Seoul, Korea: CheonJae.

Yim, J. H., Lee, D. H., Lee, Y. R., Park, S. K., & Jeong, Y. K. (2004). *Analysis and evaluation of the content relevance in the primary and secondary school mathematics* (Research Report RRC 2004-1-5). Seoul: Korea Institute of Curriculum and Evaluation.

Woo, J. H. (1992). A Korean perspective on mathematics education. *Journal of the Korea Society of Educational Studies in Mathematics*, 2(1), 119–131.

MATHEMATICS CURRICULUM STANDARDS OF CHINA

Its Process, Strategies, Outcomes and Difficulties

Xiaotian Sun
Central University for Nationalities, Beijing, Peoples Republic of China

This paper outlines the process, strategies, outcomes, and difficulties of the National Mathematics Curriculum Standards (experimental version) of China. It also presents possibilities and problems concerning the standards as well as views regarding the evaluation of the changes and prospects for the future of mathematics education in China.

THE EVOLUTION OF CURRICULUM STANDARDS IN CHINA

In order to understand the historical development of mathematics curriculum standards in China, it is necessary to understand the education system of China, as briefly introduced in Figure 5.1.

Mathematics Curriculum in Pacific Rim Coutries—China, Japan, Korea, and Singapore: Proceedings of a Conference, pp. 73–82

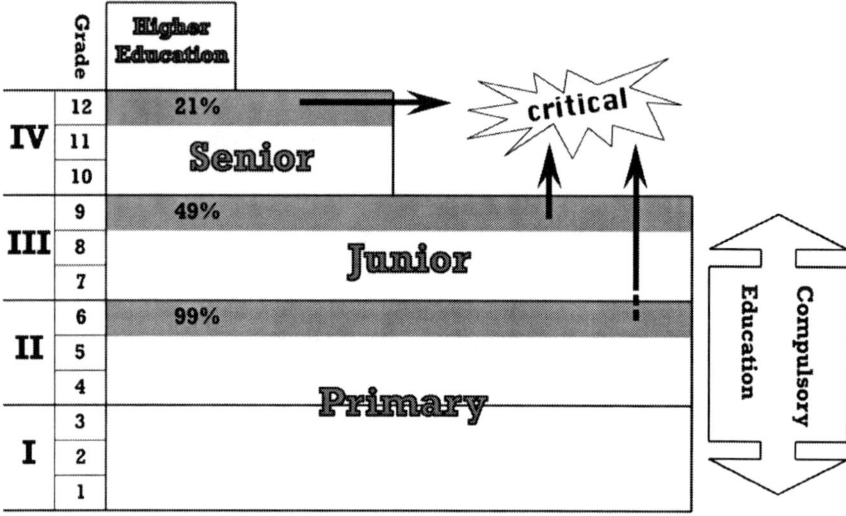

Figure 5.1.   The education system of China.

As Figure 5.1 shows, compulsory education consists mainly of primary school (1st to 6th grades) and junior secondary school (7th to 9th grades). The opportunity to enter junior secondary school is available to nearly all the students but only half of them can enter senior high. After 3 years of study in senior secondary school (10th to 12th grades) about 21% of all the students can successfully pass the entrance examination to receive higher education.

Under this system, the first framework of mathematics curriculum was issued in 1950 and revised successively in 1956 and 1963. After that framework, revisions took place quite irregularly. The 1963 version of the framework remained unaltered for 13 years due to political factors, including the Cultural Revolution (1966–1976). The 1978 version of the framework remained unaltered for 15 years during which incredible social and economical changes took place inside China. The old and out-dated framework could no longer cater to the needs of a modern educa-tion and thus hindered social and economic development. Although it was revised once by 1999, few substantial changes were brought about. The need for revision was urgent and the process was difficult. Aiming to solve this problem, in 1999 the State Ministry of Education published a guiding document *The Education Development Program for the 21st Century* (EDP). The EDP appealed for the design of a new modernized curricu-lum system and standards for primary and secondary education in China at the beginning of the twenty-first century. Only in this way can this newly developed system be implemented in 5 to 10 years in all Chinese schools. As a result, a new framework, the Mathematics Curriculum Standard (MCS) was issued for compulsory education (State Ministry of Education, 2001) and for senior secondary education (State Ministry of Education, 2003).

KEY FACTORS FOR THE NEW MCS

The birth and refinement of the new MCS largely relied on guidance from the government, the support of research projects, and the innovative ideas of the MCS working group members.

Guidance From the Government

In 2000, the State Ministry of Education issued another significant guiding document: *The National Curriculum Framework for Primary and Secondary Education* (NCF) (State Ministry of Education, 2000a). In this document, reform fields and aims for each field are elaborated at length. These reform fields include various aspects of educational change, such as curriculum aims, curriculum structure, the design of the national curriculum standards, didactical methodology, assessment, teaching materials and other curriculum resources, curriculum management, teacher training, and organization and implementation of the new curriculum. Its essential principles include:

- The process of acquiring knowledge should take into view the formation of values.

- Content should connect to a student's life and be contextually pertinent to a student's interests and customs.

- Students should be encouraged to engage in investigation, fieldwork, communication, and cooperation.

- Assessment should aim at helping students build their self-awareness and self-confidence while improving teaching.

- The adopted curriculum structure is to be synthetic, balanced, and flexible.

The major goal of the NCF is to develop students' life-long desire for learning and learning ability.

Mathematics education was the first field reformed as suggested by the guiding documents. The State Ministry of Education selected a research group in March 1999. The research group activities led to the first national mathematics curriculum standards as required by the EDP and the NCF. Only a year later the group presented these standards for compulsory students, the Mathematics Curriculum Standard (experimental version) for compulsory education (MCS 6-15). It was the first new standard among all the subjects in the national educational development program in China. In 2003 the MCS for Senior Secondary Education (MCS 16-18) was finished.

The Support of Research Projects

The following research topics have been studied as research projects since the end of the 1980s in China:

- The relation between math for science and school mathematics;
- The relationship between social needs and the math curriculum;
- The relationship between mathematics study and the healthy development of a student's body and mind;
- How students learn mathematics; and
- Understanding the trends and characteristics of international math curricula.

Most of the project's outcomes are included in a two-volume collection: *The 21st Century View for Math Education in China* (State Ministry of Education, 2000b).

The Innovative Ideas of the MCS Workgroup

For the creation of the new MCS, we should also give credit to the diligent workgroup members who contributed numerous fundamental and innovative ideas, such as:

- All students should learn mathematics, and school mathematics should be essential and appropriate to students' needs.
- The content of school mathematics should be meaningful, realistic, and challenging.
- The content should explain the processes that produce it.
- Math teaching should take into consideration a student's personal knowledge and experiences.

- Practice and independent investigation, as well as cooperation should be an important way for students to understand and grasp math.

- The math classroom should be transformed from absolutely teacher-centered to partially student-centered.

- Assessment is aimed at helping students build their self-awareness and self-confidence, while helping teachers improve their teaching.

- Math curriculum should integrate with modern informatics technology.

With these innovative ideas, the workgroup members worked to the utmost of their abilities in the refinement of the new MCS.

THE DESIGN OF THE NEW MCS

The design of the new MCS was carried out as a national research project, which was open to all interested research groups. The State Ministry of Education decides final makeup of the qualified participant groups and, if necessary, reorganizes the groups. Among all the participants, 70% were mathematics professors and university educators while 30% were mathematics teaching field practice coordinators and highly qualified classroom teachers who, in most cases, had more than 20 years teaching experience. The participants were divided into two groups to design the MCS for compulsory education and senior secondary education, respectively. The following figures illustrate the structure of the new MCS for both compulsory and senior secondary education.

In order to elaborate on the MCS for compulsory education, two examples will be employed. First is a section in the standards about "Number Sense." In the traditional Chinese curriculum only the concept of numbers and standardized calculations was explored. Number sense is a new field for our students. One of the aims for number sense is formulated as follows: "Students have to realize the diversity in the use of numbers in life. They can use numbers to represent different situations and to explain different views." The following example (p. 22, MCS 6-15) in the MCS illuminates the goal and ideas behind Number Sense for students from Grade 1 to 3:

Do a poll about "Do you like mathematics?" in your class, and use 5, 4, 3, 2, 1 representing the five different levels from "like it most" to "does not like it at all." What is your own choice? Please tell us the reason for your choice. If Xiaoming chooses 2, how does she like

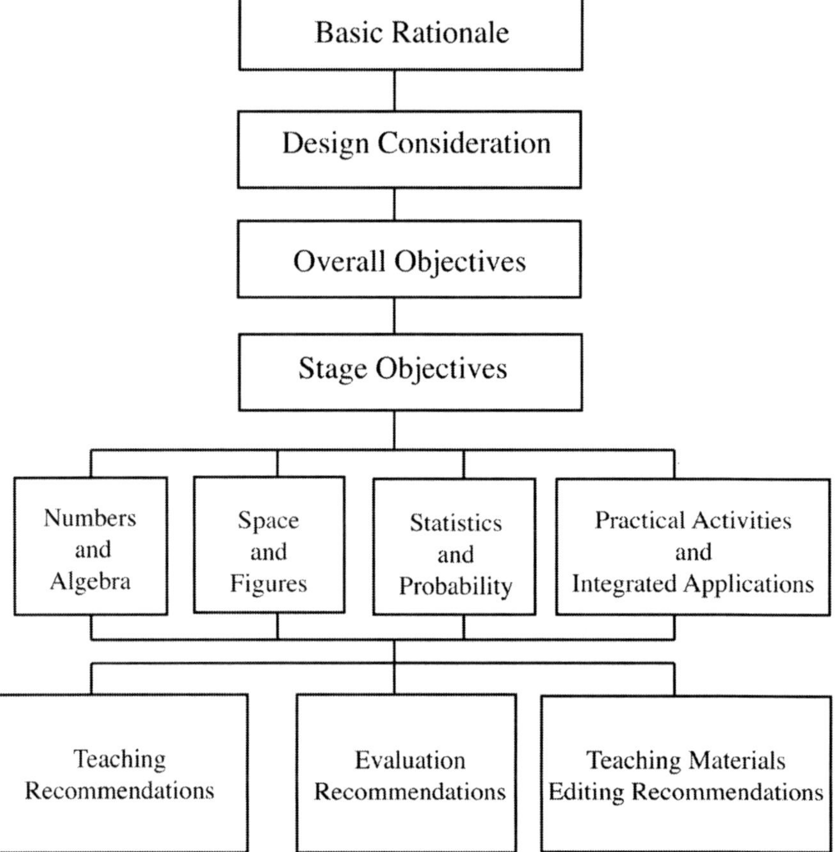

Figure 5.2.   Compulsory education.

mathematics? If Xiaoli likes mathematics, which number will she choose?

This example shows the key elements of what is meant by number sense in the newly developed curriculum. It is important that teaching and learning start in a realistic or recognizable situation. Furthermore, students are allowed to experiment, to conjecture, to reason, and to exchange views.

A second elaboration is that the new program places an emphasis on learning by means of practical activities and integrated applications. The following example from MCS for compulsory education, shows such an activity for students from Grade 4 to 6:

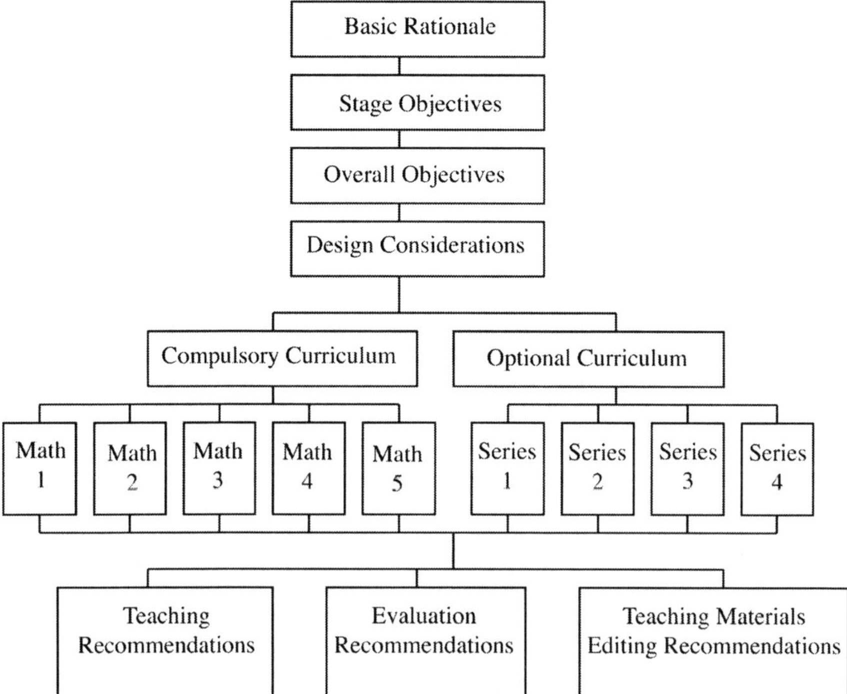

Figure 5.2.   Senior secondary education.

The TV tower problem: Try to find out the height of the Shanghai TV tower. How high is the Beijing TV tower? Approximately how many times the height of our classroom would be the height of each of these towers? How many students do you need to make a chain hand by hand that is more or less the same length as the height of each of these towers? Can you think of any other descriptions that can help us to understand how high these two towers are?

The children work together or in a group supervised by the teacher. They need to use their own experience to design strategies and solve the problem in an integrated way. They can learn how to collect data through surveys, make use of estimation and mental calculation strategies, and use different ways to measure length and height.

One of the other major changes in the mathematics curriculum is the use of a calculator. In the past, students were not allowed to take a calculator to class, because it was believed that it would limit the student's ability to make written and mental calculations. Now, because the focus has been turned to problem investigation, a calculator is a natural and indispensable tool to utilize. In MCS for compulsory education, a calculator can be used in the classroom from Grade 3 onward (9-year-old students). Moreover, there are no longer any restrictions in applying new information technologies in a classroom, if these improve both learning and teaching. MCS 6-15 also recommends developers of new textbooks to consider the use of new technologies, especially information technology.

Although the above examples show the ideas behind the standards, they only form the beginning of the desired changes. They should be regarded as results of the fundamental steps taken in development of Chinese mathematics education. The realization of new textbooks is the next step in the curriculum reform. This will also bring about a transformation of the current curriculum.

THE IMPLEMENTATION OF THE NEW MCS

Preparatory Work Before Implementation

Before the official implementation of the new MCS, meticulous preparatory work was done to obtain the most far-ranging feedback and suggestions. For instance, aiming at achieving various suggestions from as many people as possible, the MCS for compulsory education blueprint was published in eight different magazines specializing in education and mathematics, which was particularly unusual in China. Ten thousand copies of the draft version were distributed with questionnaires for further feedback. Some measures were actively taken to collect feedback from various areas. For example, the whole country was divided into nine parts geographically and from each part experts in mathematics, mathematics educators and researchers, and highly qualified teachers who have first-hand teaching experiences were invited to give this feedback. People invited also included several academic members of the China Scientific Academy, university presidents, representatives of People's Congress and 67 entrepreneurs from national enterprises. As a result, collection of the widest-ranging feedback could be facilitated. After all this preparatory work the draft was officially issued by the State Ministry of Education as a government document.

Achievement and Progress After the Implementation

Until recently, a nationally developed textbook series was only in an experimental phase. The series consisted of 21 books: six for the primary level, nine for the junior secondary level, and six for the senior secondary level. In 2001, these textbooks were used in 38 districts and the number of students involved in the experimental project accounted for about 1% of all the students in the country. Since the fall of 2005, all the primary and junior secondary schools have adopted the new textbook series for the beginning grades. As for the senior secondary students, due to the standardized entrance examination, the new textbooks were not adopted throughout the entire country. However, in five provinces, the new textbook series has been used for all the 10th-grade students (senior) and the number of provinces will be increased to 10 in the fall of 2006.

Major Difficulties in the Process of the Implementation

To implement the new MCS for compulsory education, many obstacles stand in the way. In order to adopt this completely new textbook, great efforts in teacher training are demanded. Entrenched in the traditional but out-of-date teaching methodology, most teachers are reluctant to completely accept the new teaching concepts, which often conflict with the old ones. Also, there is a dearth of additional teaching materials available for teachers to reference. Thus they chiefly rely on imitating experienced teachers while ignoring their own creation of teaching methods. This fact largely weakens one of the advantages of the new textbooks, which offer teachers more opportunities for creative teaching. Apart from inflexible teaching methods and a lack of reference materials being problematic, it has been quite difficult to imbue teachers with an appreciation for the new concepts involving testing and assessment. All these problems require persistent efforts.

QUESTIONS UNDER FIERCE DISCUSSION

There exist some questions for which there is little agreement within the development community. They generally include:

- Realistic versus formal: more or less?
- The traditional teaching system: stronger or weaker?
- The speed of implementing: quicker or slower?
- Being scrupulous (*conscientious and exact*) or being active. Can the balance be kept?
- "Two basics" (basic knowledge and skills): stronger or weaker?

We have paid much attention to reach appropriate answers, but there is still a long road ahead.

CONCLUSION

Despite these complications, constant revision of the textbook is ongoing. In April 2003, more than a third of the teachers who were using the new textbooks were surveyed and over 1,000 questionnaires collected. Between October of 2003 and June of 2004 the first revisions were carried out while the second ones started in April of 2005. Publication of the newest version is scheduled for the fall of 2006. Tenacious and meticulous efforts have been and will continue to be devoted to the perfection of the new textbooks.

REFERENCES
(ALL IN CHINESE)

State Ministry of Education. (1999). *The educational development program for the 21st century*. Beijing, China: People's Educational Press.

State Ministry of Education. (2000a). *The national curriculum reform framework for basic education*. Beijing, China: People's Educational Press.

State Ministry of Education. (2000b). *The 21st century view for math education in China*. Beijing, China: Publishing House for Beijing Normal University.

State Ministry of Education. (2001). *Full time compulsory education mathematics curriculum standard* (experimental version). Beijing: Publishing House for Beijing Normal University.

State Ministry of Education. (2003). *Ordinary senior secondary mathematics curriculum standard* (experimental version). Beijing, China: People's Educational Press.

PART II

DESIGN AND IMPLEMENTATION PERSPECTIVES ON K–12 MATHEMATICS CURRICULUM

CHAPTER 6

SIXTY YEARS OF MATHEMATICS SYLLABI AND TEXTBOOKS IN SINGAPORE (1945-2005)

Peng Yee Lee
National Institute of Education, Singapore

INTRODUCTION

Singapore is a city state with a land area of 683 square kilometers (263 square miles) and a population of more than 4 million. It is a service hub for sea, air, and finance. It was a British colony for around 150 years and became independent in 1965. The only natural resources on the island are its people. Education is regarded as an investment rather than as a social service. There are more students in the science stream than in the arts stream. Furthermore, all students have done some basic mathematics up to the Grade 10 level. This paper is an overview of important events and periods concerning teaching and learning of mathematics in the past 60 years, from 1945 to 2005. Hopefully, it can serve as background information for those who are interested in mathematics education in Singapore.

Mathematics Curriculum in Pacific Rim Coutries—China, Japan, Korea, and Singapore: Proceedings of a Conference, pp. 85–94

Since 1945, after World War II, Singapore saw a shift from the use of foreign textbooks and expatriate teachers to the use of local textbooks and locally trained teachers. In the late 1970s we saw a change from schools using many languages to instruct (Singapore has four official languages) to schools using only English as the language of instruction. However, all students continue to study their native tongue as their second language. The nature of schooling changed over the years in that it has developed a focus on providing professional training. With respect to the content of the mathematics taught, we saw the fall of Euclidean geometry and the gradual replacement of mechanics by statistics. In the last decade, we saw the shifting of classroom teaching from an emphasis on content to an emphasis on process, and from teaching to learning. These will be described in the context of Singapore.

The following five periods are covered:

- Early period from 1945 to 1959 when Singapore introduced its own syllabus;
- The implementation of the local syllabus from 1959 until a new syllabus was introduced in 1968 leading to math reforms in the 1970s;
- Math reforms and the beginning of going local;
- The back-to-basics movement; and
- The new initiatives under the banner of "Thinking Schools" and a "Learning Nation" starting in 1997.

I shall look at the K–12 mathematics curriculum mainly from a design and implementation perspective. I begin with a brief description of the educational system in Singapore during the colonial times and after Singapore's independence.

I interviewed four persons who are themselves teachers at different levels and who went to school during the times corresponding to the first four above mentioned periods. I asked the following four questions.

- What mathematics did you learn?
- What textbooks did you use?
- How did your teachers teach mathematics?
- How did you learn mathematics?

In what follows, I recorded their oral narrative. I leave it to other researchers to evaluate this oral record and to search the archives for verification.

FROM A BRITISH COLONY TO AN INDEPENDENT NATION

The population of Singapore consists mainly of migrants from China, India, and the Malay archipelago. During colonial times, there were English schools and Chinese schools, as well as Malay and Tamil primary schools for different communities. It was the English schools that produced graduates for the civil service. Other schools provided labor for the commercial sector. Though oversimplified, this does give a rough picture of the situation at the time. There was hardly any industry. Education was, and indeed still is, a gateway to a better life.

It was only after independence that mathematics and science education became dominant, in that the system came to be composed of many more science students than arts students. Education served to provide professional training for students. Eventually, education came to be considered an investment rather than a service. Students were educated to take on new challenges different from those of their teachers. They were not made merely to be replicas of their teachers as in the past.

Many traditions remain today. For example, some of the best secondary schools are boys' schools or girls' schools. Schools of different language streams with different systems finally converge into 6 years of primary, 4 years of secondary, and 2 years of preuniversity (locally called junior colleges) as it is today. Schooling is compulsory though it was not so a few years ago (Lee, 1999). Almost all students stay in school for 10 years (Grades 1–10). There are more students going to polytechnics (postsecondary institutes) than going to universities. Increasingly, more and more polytechnic graduates proceed to pursue a university degree. In Singapore, we have neighborhood schools but no village schools. The living conditions are essentially urban. It is under this setting that we proceed to describe the changes in mathematics education over the years.

EARLY DAYS

For Singapore, World War II ended in August 1945. Schools reopened in January 1946. There were simply not enough school buildings. A school might run three sessions, namely, a morning session, an afternoon session, and an evening session. This was supposed to be temporary. However, this temporary measure lasted for more than 50 years. As far back as most people can remember, Singapore schools have always had a morning session and an afternoon session. Few people remember the evening session. It was not uncommon that a village school at the time was housed in a wooden building with an *atap* roof. There could be two different grade levels of pupils in the same classroom, taught by a single teacher.

There was a Chinese school whose enrollment was 5,000. What the principal of the school said was that although the school facilities might not be very good, they were still better than the street environment. What he meant was that if he did not take students in, they would be out there in the street.

In the 1950s there were few government schools. Many English schools were run and funded by the missionaries, while the Chinese schools were run and funded by the Chinese community. In the English schools, students learned history biased toward the British and there was no Chinese language in the instruction. In the Chinese schools, they learned history biased toward the Chinese and English was a silent language (meaning that they did not learn to speak it). There was no public examination at the primary school level. In some cases, pupils sat for an entrance examination for the secondary school in which they wanted to enroll. Students from English schools sat for public examination set in England. The year 1955 was the last year that students in the Chinese stream at the senior middle three levels (Grade 12) took three examination subjects (Chinese, English, and mathematics), each written in Singapore. One year later, they were required to take six examination subjects.

In those days, there was no streaming. It was a common practice in some schools that better students would coach the weaker students outside the lessons. There was a lot of peer learning taking place. Teaching was dictated by textbooks. The syllabus meant textbook syllabus. Foreign textbooks were used. In English schools, British textbooks were used. For example, *General Mathematics* by Durell (1946) was used. In Chinese schools, Chinese textbooks were used and also American textbooks in Chinese translation, for example, *College Algebra* by Fine (1904) at the senior middle two levels (Grade 11). Kindergarten was a rarity.

Teachers came from outside Singapore. It was not uncommon that mathematics was taught by an engineer. There was a great shortage of qualified teachers. There had been cases of a Grade 7 class that was taught by a Grade 12 graduate. The class size was generally 60 students. Teachers carried a cane to the class. They were highly respected. Those were the days when not every child had the good fortune to go to school. Teaching of moral values in schools was as important as, if not more than, learning skills. As we can see, some fundamental changes have taken place over the years.

FIRST LOCAL SYLLABUS IN 1959

The first local syllabus in mathematics was drafted in 1957 and announced in 1959. It contained the mathematics syllabus from Year 1

(Grade 1) to Year 13. The syllabus adopted a spiral approach. The fierce debate at the time was whether to teach mathematics as a unified subject or with different branches; there was a choice at the time to adopt one or the other. The Chinese schools adopted the latter approach. Eventually the former approach, that is to teach mathematics as a unified subject, won out in the debate. There was no statistics in the syllabus. Calculus was taught in additional mathematics (Grades 9–10).

Schools used imported textbooks. In 1959, Nanyang University graduated its first group of students. Thereafter, many graduates of Nanyang University from the first six cohorts entered the teaching profession. Many of them became mathematics teachers. A high proportion of them became school principals. They stayed in the service until their retirement only a few years ago. They have made marked contribution to the teaching of mathematics. During the 10-year period from the 1959 syllabus, teaching would be regarded as traditional. The local syllabus was the first step toward the localization of mathematics education.

MATH REFORMS IN THE 1970S

In the 60 years of mathematics education in Singapore, the most important event is the math reforms of the 1970s. In terms of syllabus, it made a major break from the traditional content. It initiated massive retraining of teachers. Furthermore, it induced changes that followed in the next 20 years. It coincided with the forming of the Advisory Committee on Curriculum Development (ACCD) within the Ministry of Education, Singapore. The primary school mathematics syllabus was revised in 1971 and again in 1976 resulting in a reduction of content. There were emphases on numerical skills including approximation. The secondary school mathematics syllabus, known as Syllabus C, was introduced in 1968 and revised in 1973. The so-called modern mathematics, such as sets, was introduced at the Secondary One (Grade 7) level. Teaching of Euclidean geometry was drastically curtailed.

In addition, statistics was provided as an alternative to mechanics. However, due to the shortage of teachers, there were still more schools teaching mechanics than teaching statistics. Number bond had already entered primary curriculum. Algebra was introduced at the Primary 5 and Primary 6 levels (Grades 5–6). Calculus was taught, but only at junior colleges (Grades 11–12). The first set of locally produced secondary school textbooks appeared (Teh, 1969). Other locally produced textbooks followed.

The teaching approach had changed somewhat. For example, worksheets were used. Some teachers used Cuisenaire rods to teach four operations of the whole numbers at the primary level. However the discipline remained strict. Pupils were caned (on hands only) for failing their tests. There was no homework for primary schools. There was more peer learning than direct teaching from teachers. Streaming of students had not been introduced.

The landmarks during the period were:

- Elementary mathematics at the O level (Grades 7–10) was made compulsory as an examination subject as of 1974.
- From 1974 to 1977 more secondary schools switched to Syllabus C and the failure rate increased.
- Science and mathematics were taught and examined in English starting from 1976 but it was not fully implemented until 1984.

In conclusion, though the math reforms were a worldwide phenomenon, they had their local implications. They hastened the localization of syllabus and textbooks. They led to the production of CDIS (Curriculum Development Institute of Singapore) primary textbooks in the 1980s.

BACK TO BASICS

The math reforms led to the falling of mathematics standards. In particular, students became weak in algebraic manipulation. There was a call of *back to basics* by the mathematics community worldwide. Singapore followed the trend. It also coincided with the implementation of the New Education System (NES) in 1979. Under the NES, students were streamed into two and later three streams. The objective of streaming was to reduce the attrition rates in the school system. Each stream then had its own syllabus. New syllabi were adopted in 1981. In particular, for the secondary schools (Grades 7–10) Syllabus C phased out progressively and was replaced by Syllabus D. As far as content is concerned, the syllabus has not changed much since then. The current secondary school syllabus is in a sense Syllabus D plus. In other words, it is a modified version of Syllabus D. In Syllabus D those so-called modern topics were moved out of secondary one and two levels (Grades 7–8). Some disappeared and some were kept in secondary four (Grade 10). One rationale was to make sure that students leaving schools at the end of secondary two (Grade 8) would have enough trigonometry to learn a trade and to work in the maritime industry. In such cases, modern maths were not helpful.

Also under the NES, schools using various languages as their medium of instruction gradually evolved into a single system using only English. During the changeover some pupils learned their multiplication tables half in Chinese and the other half in English. Mathematics teachers were retrained to teach the subject in a different language.

Concerning textbooks, the Ministry of Education set up a special unit, called CDIS as mentioned earlier, to publish textbooks. The primary school textbooks adopted in the United States were produced by CDIS during the period. The popular secondary school textbooks were published commercially. Primary schools tend to follow textbooks closely, though this is not necessarily so for secondary schools.

Scientific calculators were introduced into secondary schools (Grades 7–10) starting in 1982. Initially they were introduced in secondary two (Grade 8), later they were introduced in secondary one (Grade 7). On the whole we were cautious in adopting the use of calculators. For example, graphic calculators will be introduced into junior colleges (Grades 11–12) only in 2006, and we have not yet introduced calculators into primary schools.

Teaching remained relatively traditional. For example, a 0 mark would be awarded if the answer is wrong with no reference to the process. Surprising or perhaps not so surprising, peer learning among students was common even though we seem to have the impression that there was more teaching than learning at the time.

In summary, we practiced differentiated teaching with more than one syllabus. Mathematics was taught and assessed in English, not necessarily the home language of the student. We did not change that much during the math reforms as compared with our neighboring countries. Therefore, we also had less to reverse. By the early 1990s, we had reached a kind of equilibrium state for syllabi and textbooks. Both were designed locally. In what followed, the changes were mainly, though not exclusively, regarding how to teach mathematics rather than what to teach.

THINKING SCHOOLS AND A LEARNING NATION

This idea was put forward in 1997 by the Ministry of Education. It started a new era of educational reform with which we are presently engaged. Some of the events and the rationale behind those events are explained in another conference paper by Soh (see chapter 2). The three initiatives are national education, information technology and thinking. By national education, we refer to good citizenship and nation building.

For information technology, we have gone through two master plans. In Master Plan I (1997–2002), computers were brought into schools and teachers were trained to use them. In Master Plan II (2002 – 2008), there is to be integration of information technology into the curriculum. Also, the introduction of information technology will induce changes in teaching content and assessment. Furthermore, it should help students with learning difficulties to overcome handicaps in basic mathematical operations (Ang & Lee, 2005).

By "thinking," we refer to the world-wide trend of more learning and less teaching. As a consequence, we reduced content by at least 10%, with a larger reduction in the secondary schools and less in the primary schools. Hopefully, students can spend more time with mathematical process, and realize mathematics is not just a collection of formulas. A learning nation refers to adult education outside the school environment, making learning a life-long endeavor.

In summary, the 60 years can be divided into five periods as shown below. For easy reference, figures are rounded.

- 1945–1960: Early days;
- 1960–1970: First local syllabus;
- 1970–1980: Math reforms;
- 1980–1995: Back to basics; and
- 1995–2005: New initiatives.

Over the years the major events include:

- Merging of different language schools;
- Streaming of students with different abilities;
- Changing foreign syllabi and textbooks to those locally produced;
- Shifting from expatriate teachers to locally trained teachers; and
- Changing of emphasis from professional training to nurturing entrepreneurship.

These major events led to the following outcomes:

- Change in language of instruction to English;
- Shift to practice differentiated teaching;
- Change in streaming: there are more science students and mathematics is made compulsory;
- Build up of local expertise in textbook writing, teacher training, and research; and
- Develop our own system since there is nowhere to copy.

In the years to come, we have to find our own solutions to our unique problems. Our approach will be pragmatic. We shall do what will work for us.

TEACH LESS TO MORE

This is my personal view. By "teach less to more," I mean "teach less mathematics to more students." This is certainly one way if not the only way to create a bigger pool of mathematics students. Since jobs are now mostly in the service industry, we must realize that the service industry requires less mathematics than the manufacturing industry. So we may have fewer and fewer students doing mathematics. If we have a bigger pool, more will excel and some will become mathematicians. Some will become schoolteachers, so that there is a hope that mathematics will be taught well and our students will learn mathematics well.

During the colonial times, education was essentially to train civil servants for the colonial government. For more than 3 decades after the independence of the Republic, the focus of education in Singapore became primarily to provide engineers and managers for the multinational companies. Nowadays, the role of education, in addition to providing manpower for the employment market, is to nurture entrepreneurs so as to increase the wealth of the nation. Correspondingly, we have to teach mathematics differently.

Looking forward, we make a list of what we think might happen or what we would like to see happen.

- There will be a genuine reduction of content.
- We shall make thinking more explicit in teaching.
- There will be other modes of assessment.
- The use of technology will make mathematics learning more experimental.
- We expect an increasing use of resource materials in teaching.

The above items are all interrelated. For example, the reduction of content will give more time for learning mathematics differently.

The production of suitable textbooks is a real challenge. The Ministry of Education ceased to publish textbooks. It was left to the commercial publishers to tender for it. Hopefully, through competition, better textbooks are produced, and also if changes are necessary, a private enterprise would probably move faster than a government department.

Assume that we have a good syllabus. Assume that we have good textbooks. Assume that the high-stake examinations have been downplayed. We still need teachers. This is probably the most important factor in the entire business of mathematics education. Teacher preparation is the key to a successful implementation of an intended syllabus (Lim-Teo & Suat, 2002). The difficult question is how to train teachers to teach not only for examinations but also for learning. Teachers are under great pressure from principals and parents to produce good examination results. Yet it is also the professional duty of teachers to make sure that students learn enough for the next stage of their studies. We say examination drives learning. We often say it in a negative sense. However, it could also be positive. Sometimes it can even be an effective way to implement a new initiative, that is, to do it through the examination. In fact, some open problems have been set for the Primary School Leaving Examination in Singapore.

As a final point, the teaching and learning of mathematics in Singapore schools is regarded as more structured and more officially regulated compared to many other countries. This will remain so for at least the next 5 years. However, there will be more room for individual innovation, especially in classroom teaching. This is happening now, and in times to come, it will grow. How fast it will grow depends on how effective the training of teachers will be and on the availability of local resource materials for teachers (see, for example, Lee, 2005, 2006).

REFERENCES

Ang K. C., & Lee, P. Y. (2005). Technology and teaching and learning of mathematics—The Singapore experience. *Asian technology conference in mathematics (ATCM) 10th anniversary CD-ROM*. Blacksburg, VA: Advanced Technology Council in Mathematics

Durell, C. V. (1946). *General mathematics*. London: George Bell & Sons.

Fine, H. B. (1904). *College algebra*. Boston: Ginn & Company.

Lee, P. Y. (1999). Mathematics education in Singapore. In Z. Usiskin (Ed.), *Developments in mathematics education around the world: Proceedings of the fourth UCSMP international conference on mathematics education* (pp. 188–193). Reston, VA: National Council of Teachers of Mathematics.

Lee, P. Y. (Ed.). (2005). *Teaching secondary school mathematics: A resource book*. Singapore: McGraw Hill.

Lee, P. Y. (Ed.). (2006). *Teaching primary school mathematics: A resource Book*. Singapore: McGraw Hill.

Lim-Teo, S. K. (2002). Pre-service preparation of mathematics teachers in the Singapore education system. *International Journal of Educational Research, 37*, 131–143.

Soh, C. K. (2005, November). *An overview of mathematics education in Singapore*. Paper presented at the First International Mathematics Curriculum Conference, Chicago.

Teh H. H. (Ed.). (1969). *Modern mathematics for secondary schools*. Singapore: Pan Asia.

CHAPTER 7

DESIGN AND IMPLEMENTATION OF KOREAN MATHEMATICS TEXTBOOKS

JeongSuk Pang
Korea National University of Education

INTRODUCTION

Whereas educational leaders in Korea have recently attempted to provide for some degree of autonomy at the local school level, curricular materials such as mathematics textbooks and teacher guidebooks are very influential, leading to directive, coherent, and rather uniform changes. Specifically, as almost all Korean teachers use textbooks as their main instructional resources, the development of good textbooks has been an important concern in shaping K-12 Korean mathematics education. This is especially true for elementary mathematics education, because Koreans use only one elementary mathematics textbook series. In other words, all Korean elementary school students study mathematics with the same textbooks, which consequently serve as the bottom line for teaching and learning.

Mathematics Curriculum in Pacific Rim Coutries—China, Japan, Korea, and Singapore: Proceedings of a Conference, pp. 95–125

The design and implementation of instructional materials are closely linked to changes in the national mathematics curriculum. In fact, Korean mathematics education reform centers around revising the national curriculum in concert with textbooks and teacher guidebooks. It is of great significance to explore the processes whereby principles and expectations concerning school mathematics come to influence textbook development.

To pursue these interests, this paper first provides an overview of textbook development, introducing basic information about the transition process from the national curriculum to textbooks, types of textbooks, and textbook authors. This paper then presents more detailed information about writing elementary and secondary mathematics textbooks in terms of the general flow of textbook production, contributing factors, mathematical topics, and the revision or approval processes. Next, the paper deals with how to select and use textbooks at the local school level. Finally, this paper reviews several issues and concerns that are under discussion on the future development of mathematics textbooks.

OVERVIEW OF TEXTBOOK DEVELOPMENT

Guide: Transition from Curriculum to Textbook

Mathematics textbooks are developed to follow and specify what the national curriculum intends. The most recently developed seventh curriculum has a level-based differentiated structure, and emphasizes students' active learning activities in order to promote their mathematical power, namely, problem-solving ability, reasoning ability, communication skills, connections, and dispositions (Ministry of Education, 1998). This curriculum resulted from the repeated observation that previous curricula were rather skill-oriented and fragmentary (especially in conjunction with the expository method of instruction), and that previous curricula did not consider differences among students with regard to mathematical abilities, needs, and interests (Lew, 1999). Thus, the main motivations for the current curriculum are increasing concern for individual differences and the desire to provide maximum growth of individual students on the basis of their abilities and needs. Given this curricular intent, mathematics textbooks should provide students with opportunities to nurture their own self-directed learning and to improve their creativity.

To promote these outcomes, the Ministry of Education and Human Resources Development (MOE) publishes a guide for writing elementary and secondary mathematics textbooks. The guide first provides general directions for writing textbooks and explains the current curriculum in terms of goals, characteristics, principles, emphases at each school level, and so on. The guide then introduces very specific cautionary notes for what to include, and what to exclude, how to structure the textbooks, and how to choose and present mathematical contents, as well as other factors. As well, the guide specifies the physical style of textbooks such as size, quality of paper, color, sequence of editing, length, font size, and so on.

The following general guidelines apply to all grade levels. First, textbooks should select mathematical content that individual students can use to improve their creative thinking and reasoning ability. At some point in a learning sequence, instructional resources should be presented differently for students of different mathematical attainment. Whereas high-achieving students confront advanced tasks including real-life complex situations, their low-achieving counterparts solve basic problems involving fundamental understanding of important mathematical concepts and principles.

Second, textbooks should contribute to improving the process of teaching and learning. Most importantly, textbooks have to underline a learning process by which students solve problems for themselves through individual exploration, small-group cooperative activities, or discussion.

Third, textbooks are to be easy, interesting, and convenient to follow on the part of students. For instance, instructions of games or activities in the textbooks should be specific enough for students to initiate them without a teacher's further explanation and demonstration. Textbooks should take into account students' various interests and stimulate their motivation to learn. They should be designed to promote readability among students. Textbooks should incorporate appropriate use of different multimedia learning resources.

Fourth, textbooks are to be flexible in a way that teachers can refine or even revise them, as they reflect on the characteristics of their schools or provinces. It should be recognized not as the sole material to be followed but as an illustration embodying the idea of the curriculum.

The guide also provides a comparison between a recommended textbook and a traditional one in terms of six aspects in order to give a clear idea of what kinds of textbooks are expected (see Table 7.1).

For emphasis, the guide illustrates how to present mathematical contents from the national curriculum. For example, for the textbooks of eighth grade, the following are included:

- With regard to "repeating decimals," you may deal with converting a fraction into a repeating decimal, and vice versa, but do not deal with basic operations among repeating decimals.
- With regard to "division of polynomial expression," you treat only the case that the quotient of dividing a polynomial by a monomial expression is a polynomial.

Table 7.1.   Recommended Textbooks Versus Traditional Textbooks

	Traditional Textbook	*Recommended Textbook*
Perspectives of textbook	• The sole and most significant textbook • Mathematics education depends on the textbook • Textbook mainly focuses on knowledge	• Main but one among various instructional resources • Mathematics education depends on curriculum • Textbook concerns not only knowledge but also skills and attitudes
Statements in textbook	• Summarizes knowledge, condenses concepts, lists the essential points for lecture	• Presents various facts, provides specific cases, deliberates the process of learning (procedure and method)
Structure of a unit	• The one and only structure applied to all textbooks	• Various structures depending on the characteristics of given topics
Selection of contents	• Knowledge-based contents • Teacher-based contents • Minor connections to real-life contexts	• Real-life experience related to important concepts • Case-based, student-based contents • Consideration of utility
Construction of contents	• Linear construction in terms of mathematical structure • Monotonous construction of sentences and illustrations	• Nonlinear construction considering relevant knowledge and real-life experience • Various designs and editing
Process of research and development	• Textbook development with minor basic research	• Textbook development based on basic research

- With regard to "inequality," you may focus on the properties and solution processes of an inequality. Therefore, you have to avoid an inequality with complicated or letter coefficients.

- In addressing a graph of $y = ax + b$, you have to adjust a and b in a way that students can easily draw a graph.

Types of Textbooks

There are three models for development of different kinds of Korean textbooks (Ministry of Education and Human Resources Development, 2000). For the first, the Education Ministry consigns textbook development to an appropriate university or research institute. This is typical of most elementary school textbooks including mathematics textbooks, special education textbooks, and a few secondary school textbooks, dealing with Korean language, history, and ethics. Probability and statistics as well as discrete mathematics for 11th and 12th graders are also of this type. For the second model, multiple teams consisting of university-based educators and in-service teachers write the textbooks, which are assessed and approved by the ministry. This case is typical of secondary school textbooks including mathematics textbooks. In the third model, noneducators write textbooks, which are to be adopted by the ministry. This is typical for some optional subject matter textbooks at the high school level such as computer education and religion.

Consequently, whereas there is one elementary mathematics textbook series, there are various secondary counterparts. Fully, 16 mathematics textbooks are currently used for seven to 10 grades. For 11th and 12th graders who choose from among a range of mathematics subjects, there are 12 mathematics I and mathematics II, 8 calculus, 4 practical mathematics, and 1 probability and statistics and discrete mathematics textbooks. Despite different sorts of secondary mathematics textbooks, the main contents are identical because textbook authors need to follow the specific guidelines described above. Given that the general form and design of textbooks are predetermined by the government, textbook authors attempt to polish their books, for instance, by linking mathematical concepts or principles to be addressed into real-life situations, emphasizing an inquiry-based approach in learning mathematics, or providing reading materials and performance tasks for further self-directed learning.

Textbook Authors

Construction of a Team

Since there is only one elementary mathematics textbook series, the Ministry pays careful attention to select an appropriate research and development institute and to appoint a principal researcher who leads the overall process of textbook development. Once an institute is determined, the principal investigator constructs both a research and a writing team. The research and the writing team of the institute must consist of teacher educators, practicing teachers with at least 5 years' teaching experience, and senior researchers with at least 5 years' research experience. The government, superintendents of local school districts, or university professors recommend particular in-service teachers. The principal researcher of the appointed institute plays the most important role in determining who participates in the writing team.

The current elementary mathematics textbook research team consists of about 10 people who are mainly professors of mathematics education at national universities specializing in elementary education, and two or three people from the ministry (Ministry of Education and Human Resources Development, 2001). The writing team consists of eight to 10 people for each mathematics book. Whereas about three people of the writing team are professors of elementary mathematics education, the others are in-service teachers mainly selected by the professors. Professors often author multiple units at different grades, teachers usually write for only one or infrequently two units. In fact, more than 50 elementary school teachers have been involved in authoring the current elementary mathematics textbooks. In an effort to make writers fully responsible for the quality of textbook contents, each textbook includes the name of the corresponding author of each unit.

To be clear, the ministry plays an important role in selecting a research and development institute as well as monitoring the overall process of textbook development, but does not have a direct connection with individual authors. Textbook authors also have no direct connection with a publishing company. They usually work within the framework constituted by the institute.

As secondary mathematics textbooks are approved by the government, many people volunteer to write textbooks. Authors of elementary mathematics textbooks work for the government, those of secondary counterparts work for commercial publishers. Usually, around three to eight university faculty members in conjunction with in-service teachers constitute a writing team. A general tendency is that mathematics professors are involved in writing secondary mathematics textbooks more so than mathematics education professors. In-service teachers included in a writing team usually are mathematics education graduates from the same university or graduate school. More than ever, many in-service teachers have been involved in writing secondary mathematics textbooks. Teachers' active participation in authoring textbooks is a new trend.

Characteristics

The collaboration between university-based researchers and school-based teachers reflects a negotiation between theory and practice. On the one hand, the textbook should be based on thorough research concerning the problems of current textbooks, the principles and directions of the updated national curriculum, the various perspectives of teachers, parents, and students, the accumulated results on how students learn mathematics, and other factors On the other hand, the textbook should be customized in a way to reflect the needs and realities of mathematics classrooms. For this reason, many experienced teachers are involved in writing textbooks.

Textbook authors are not granted time away from their regular jobs to write textbooks. As they have to do their regular duties, being textbook authors entails a substantial commitment. Nevertheless, many professors and teachers are willing to be textbook authors. They receive only a small amount of money or a royalty for their manuscripts. In addition, the authors of secondary mathematics textbooks may have an opportunity to earn money by writing corresponding reference books or workbooks for students. But the inducement is related more to the honor of being a textbook author than to any direct economic reward. Given the great importance of mathematics textbooks in the Korean context, just being an author is an honor by itself.

WRITING OF TEXTBOOKS

General Flow of Textbook Development

Since the current mathematics curriculum is different from the previous ones, it is important that its goals be reflected in the new textbooks. The recent mathematics textbooks were developed sequentially. Table 7.2 shows a rough outline of textbook development and implementation over a series of years.

Elementary Mathematics Textbooks

As noted, the characteristics and processes of developing elementary mathematics textbooks and secondary ones are different. Figure 7.1 shows a general flow of developing elementary textbooks. The MOE announces a national mathematics curriculum and establishes an overall plan for developing instructional materials such as textbooks, workbooks, and teacher guidebooks. The MOE then selects a research and development institute and trusts the institute with the development of textbooks and related resources. The assigned institute submits a proposal concerning the development, which is reviewed and finally approved by the government.

The institute organizes both a research and a writing team. The research team consists of four or five professors from the institute, one or two professors from other institutes specialized in elementary education, and two ministry officials. Whereas a writing team operates per grade, the research team works across grades. The main role of the research team is to give a concrete blueprint to the national curriculum so that the basic principles of curriculum are specified in the textbooks. Given this, the institute sets the basic directions of writing textbooks and specifies what to do, which are reviewed again by the ministry. At this time, the ministry

Table 7.2. Outline of Textbook Development and Implementation

Grade	1997	1998	1999	2000	2001	2002	2003	2004
1st, 2nd	Announcement of national mathematics curriculum	Development (D)	Experiment (E)	Implementation (I)				
3rd, 4th			D	E	I			
5th, 6th				D	E	I		
7th		Announcement of Approval Plan	D	Approval (A)	I			
8th				D	A	I		
9th					D	A	I	
10th			Announcement of Approval Plan	D	A	I		
11th					D	A	I	
12th					D	A		I

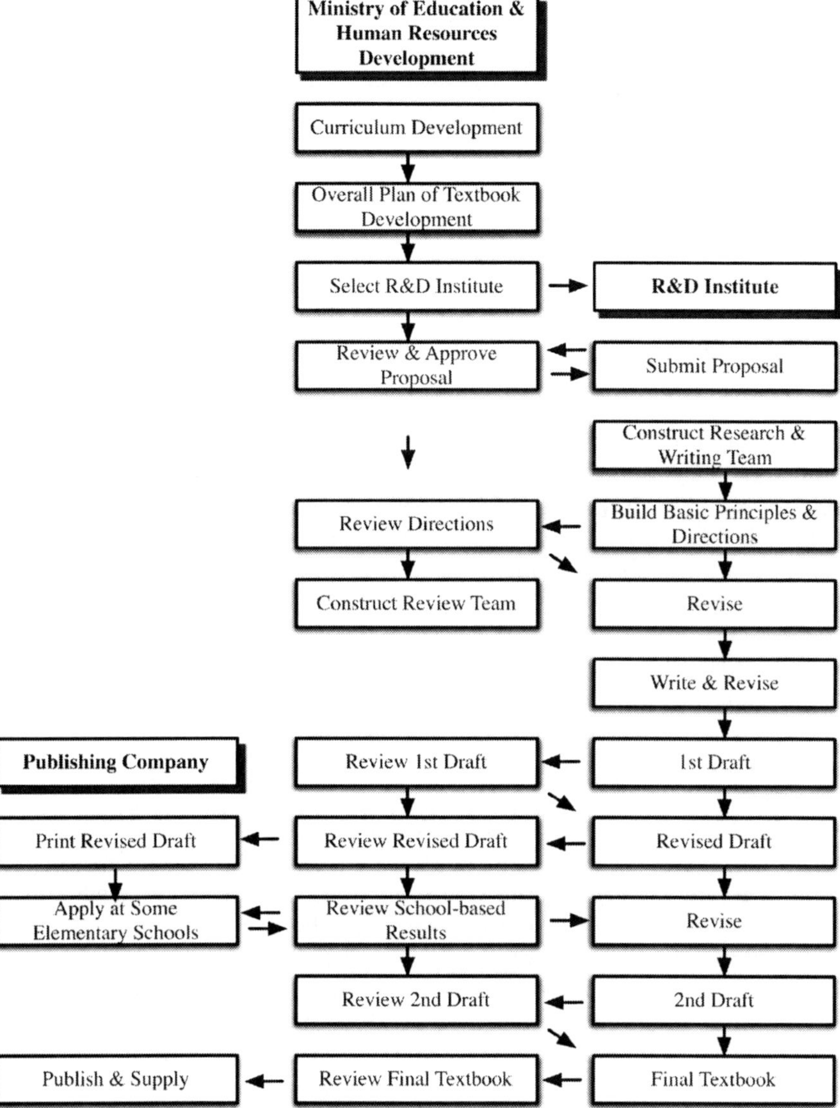

Figure 7.1.   General flow of elementary mathematics textbook development.

constructs a review and assessment board of textbooks in which about 10 professors serving for elementary mathematics education and about 10 teachers or supervisors from elementary schools and/or local educational offices across the country are joined.

The writing team of the institute creates manuscripts and goes through revisions as needed. On the basis of consultation with the research team, the first draft is prepared. The review and assessment board examines the draft and asks for revisions. The institute revises the draft and the board reviews it again. The revised textbook is applied to some elementary schools for 1 year in order to diagnose the strengths and weaknesses of the textbook (see the revision section for details). The writing team prepares a second draft, drawing heavily on an analysis of school-based results, which is again reviewed by the board. In this way, the final version of a textbook is made and published. Table 7.3 shows a time line of this whole process with the example of developing textbooks for third and fourth graders (Ministry of Education and Human Resources Development, 2000).

Secondary Mathematics Textbooks

Figure 7.2 shows a general flow of developing secondary mathematics textbooks. The MOE announces a national mathematics curriculum and establishes not only an overall plan for developing instructional materials, but also a basic approval plan, including guidelines for textbook writers and criteria of approval.

The MOE then empowers an approval committee of more than 20 members. It is the Korea Institute of Curriculum and Evaluation (KICE) that establishes a detailed plan for approval procedure. Publishing companies apply for approval. In order to participate in textbook development, publishers must have more than 20 related books or instructional materials.

A publishing company submits its first draft of textbook to KICE. The approval committee evaluates the draft and asks for revisions as needed. The company then resubmits a second draft and the committee checks whether or not the requested revisions have been completed. If the textbook turns out to be satisfactory, the company prepares the teacher guidebook to be used with the textbook. Once the company submits the guidebook, the committee assesses it. The MOE finally announces textbooks approved by the committee. It usually takes 7-8 months from submitting the first draft to getting approved. Approved textbooks are published and supplied under the auspices of the Association of Korean Textbooks.

Contributing Factors

Influence of Research: Selection and Presentation of Contents

The degree to which research influences the selection and presentation of lessons is rather limited and indirect. To be clear, there are two main

Table 7.3.   Time Line of
Textbook Development for 3rd and 4th Grades

Process	Division of Works		Time	
	MEHRD	*Institute*	*1st Semester*	*2nd Semester*
Decision of institute	o		April 1998	April 1998
Establishment of criteria & guidelines for writing textbooks	o		May 1998	May 1998
Writing & submitting a R & D proposal		o	May 1998	May 1998
Review & approval of the proposal	o		May 1998	May 1998
Writing & submitting a detailed proposal (e.g., analysis of problems, directions of reflecting curriculum)		o	October 1998	October 1998
Constructing a review committee & deciding details of writing textbooks	o		May– December 1998	May– December 1998
Writing textbooks & producing pictures and illustrations		o	November 1998– September 1999	November 1998– September 1999
Submitting a mid-report		o	December 1998	December 1998
Reviewing a mid-report	o		December 1998	December 1998
Submitting a 1st draft		o	June 1999	February 2000
Reviewing a 1st draft	o		August 1999	March 2000
Constructing, submitting, & reviewing a revised draft	o	o	September 1999– October 1999	April 2000– May 2000
Ready for school experiment (print and publish)	o		December 1999– February 2000	June– August 2000
School experiment	o	o	March–June 2000	September 2000
Reflecting school experiment		o	August 2000	February 2001
Constructing and submitting 2nd draft		o	September 2000	March 2001
Reviewing 2nd draft	o		October 2000	April 2001
Constructing final textbooks		o	November 2000	May 2001
Implementation	o		March 2001	September 2001

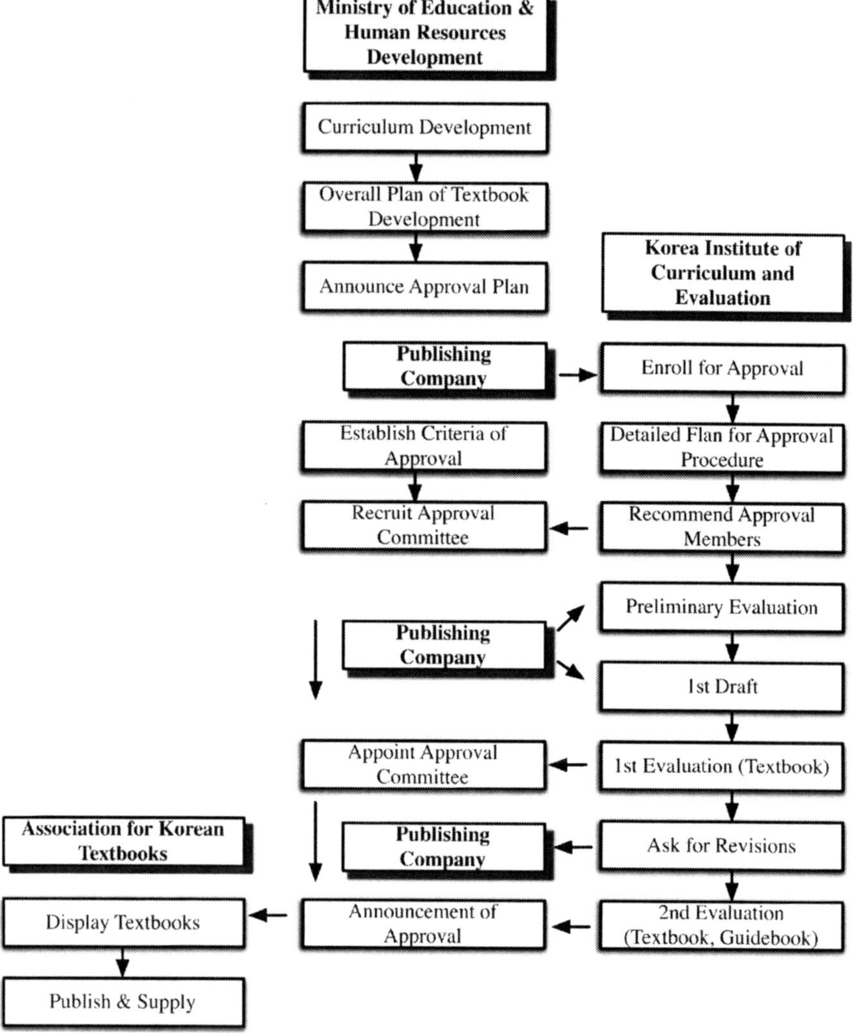

Figure 7.2.   General flow of secondary mathematics textbook development.

institutes in which various studies on textbooks are conducted. One is Korea Textbook Research Foundation (KTRF) and the other is KICE. KTRF attempts to improve the quality of Korean textbooks by conducting overall studies on textbooks such as characteristics and functions of textbooks, cross-national comparative analyses of textbook policy, textbook use and evaluation in schools, systems of textbook development, establishment of constant textbook revision system, and the like.

Whereas KTRF deals with overall aspects of textbook development and policy, KICE focuses more on subject-specific textbook research. For instance, KICE deals with development of a model of mathematics textbooks, national assessment of mathematics achievement, implementation of current mathematics curriculum at elementary and secondary schools, development of a mathematics teacher training program model for improvement of classrooms, development of mathematics instructional materials, and so on. In addition to the products of the two institutes, there are many individual studies that may be related to issues of selecting and presenting mathematics lessons. Such studies include a comparative analysis of Korean textbooks and those of other countries (e.g., Kim, 1999; Park & Yim, 2002).

However, it is up to textbook writers whether such research has a direct influence on textbook development. There is no obligation for textbook writers to reflect current research on mathematics education within Korea or abroad. In the case of elementary mathematics, teacher guidebooks include general descriptions of the characteristics and directions of elementary mathematics education, instructional methods, and lesson models in which several references are mentioned. However, research interests and issues of the principal writers tend to be the key elements. The current methods of presenting mathematics lessons reflect the principal writers' philosophy of elementary mathematics education (Bae, 2004).

In the case of secondary mathematics, many teachers' guidebooks state that recent mathematics education trends and studies have influenced their textbook development. Such guidebooks often summarize the characteristics of mathematics curricula of other countries such as the United States, Japan, and the United Kingdom. Some guidebooks also describe overall learning theories of Bruner, Skemp, Dienes, Ausubel, Gagne, Freudenthal, and Polya (e.g., Woo, Lew, Moon, & Park, 2002). Besides general references in and out of Korea, some textbook writers consult textbooks from other countries, mainly to find information on relating mathematical concepts or principles to everyday life, or for insights into presenting mathematical ideas. In short, it depends on the textbook writers as to whether and how much research influences the selection and presentation of mathematics lessons.

Influence of Research: Grade Placement

Research influences the grade placement of mathematical topics as we develop our national mathematical curriculum. The mathematics curriculum describes which topics and their related contents need to be covered at which grade. Research about the grade placement of topics is conducted both by individual researchers and by KICE. Individual research includes cross-national comparative study of specific mathematical concepts (e.g., at which grade a fraction is introduced and how it is presented as grades go up), as well as surveys of major mathematical areas for specific grade levels (e.g., how sixth graders understand ratio). These studies are conducted sporadically, mainly by professors and in-service teachers.

KICE conducts research more systematically and broadly. Whenever it is time to change our national curriculum, KICE surveys whether mathematical topics and contents are appropriate at each grade level (Yim, Lee, Lee, Park, & Jeong, 2004). In the survey, elementary and secondary school students evaluate the amount, difficulty level, and interest of contents they have to learn at each grade. Teachers and professors (including textbook writers) evaluate the amount of learning (e.g., number and depth of learning themes; number and difficulty level of textbook problems and exercises), level of significance and interest of main topics, validity and appropriateness of approaches (e.g., whether textbook contents are effective for achieving the educational purposes reflected on the curriculum; whether each content is necessary at specific grade level), and interconnections among contents (e.g., whether there are contents that lack connections within or across units).

In particular, teachers from first to 12th grade are asked to evaluate whether each unit of their textbooks needs to be presented as it is, deleted or moved to another grade level, or otherwise requires increased or decreased attention. They are also expected to provide some rationale or evidence for their evaluation. For instance, it is reported that second grade students experience difficulties in the applications of the four basic operations and in the understanding of terms such as slide, flip, and turn. In addition, it is reported that textbook activities of spatial sense are fun but mathematically weak and that contents of patterns are presented unsystematically. In the case of seventh grade, whereas spatial sense and the use of various examples and models are recommended for increased attention, other areas are recommended for decreased attention such as the concept of sets, applications of greatest common divisors and least common multiples, and the base two system.

In addition to the survey, since 2000 KICE has conducted the National Assessment of Educational Achievement to monitor the quality of education and appropriateness of the national curriculum as well as to provide background information on achievement levels to students, teachers, parents, and government. Students' achievement levels influence the grade placement of mathematical topics.

To put it all together, educators who are in charge of developing the national mathematics curriculum determine topics and contents specific to each grade level on the basis of various individual studies, surveys of teachers, students' achievement, and comparative analysis of our and other countries' curricula. Textbook writers do not have authority to change the grade placement of topics.

Construction of Contents and Problems

Nonrepeating Structure

Our mathematics textbooks have a linear structure so that each topic presented at a certain grade is expected to be mastered at that grade level (Grow-Maienza, Beal, & Randolph, 2003). We do not repeat the same content across grades. Our textbooks emphasize internal connections of similar mathematical topics within a grade and across grades (Pang, 2004). However, this does not mean that the same topic is presented repeatedly and spirally. In fact, the guideline for textbook writers includes the idea that units at each grade level should be balanced, and unnecessary repetitions or illogical development should be avoided.

Textbook Versus Workbook

Our textbooks provide students with various practice activities including basic problems, application problems, performance assessments, and/or projects. However, such textbook activities are limited in terms of the number, type, and difficulty level of problems. In the case of elementary mathematics, usually simple problems are introduced in textbooks, whereas more difficult ones are presented in workbooks. In fact, each elementary mathematics textbook has its concomitant workbook. Whilst the textbook centers around mathematical activities and thinking processes by which students can learn mathematical concepts and principles, the workbook plays a major role in reinforcing such concepts by letting students solve various problems. Textbooks are employed mainly in actual classroom instructions, whereas workbooks are usually used for students' self-practice or homework.

Occasionally, teachers are expected to conduct unit assessments in workbooks. Unit assessment consists of two parts. The first part is the same for all students, providing an overall examination of understanding mathematical ideas of a given unit. The second part is tailored to each individual on the basis of the result of the first part.

In the case of secondary mathematics, each textbook-publishing company publishes a students' self-learning book that includes a concise summary of textbook contents, detailed solutions to problems in the textbook, various reading materials, and many problems to prepare for middle and final examinations. While elementary school teachers use mathematics workbooks as part of their instructional materials, their secondary counterparts usually do not rely on students' self-learning books. Textbook writers do attempt to include meaningful and varied problems within their limited page allotment, but at the same time they assume that most students take advantage of self-learning books to anchor key skills and understanding.

Revisions and Approval of Textbooks

As described, the development of a new textbook and its related instructional resource is of great importance in shaping Korean mathematics education. Therefore, we make every effort to develop a textbook that can foster students' mathematical power as much as possible. Many steps are involved in a systematic and continuous review process leading up to the final version: a pilot experiment at a local elementary school level; discussion and consultation among experts with various backgrounds in mathematics or mathematics education, and so forth.

Revisions of Elementary Mathematics Textbooks

In addition to the writing team's ongoing revisions, approximately four formal revisions are made before a completely new series of textbooks is initiated (MOE, 2000). Once the writing team submits their first draft, the review and assessment board examines the draft and asks for revisions. The writing team then revises the draft (first revision) and resubmits the revised draft. The board reviews the draft again and the writing team prepares a draft for the elementary schools designated for the pilot examination (second revision). The writing team then revises the textbook on the basis of an analysis of school based results (third revision). This is again reviewed by the board, which suggests the revisions needed to finalize the textbook (fourth revision). Despite various comments from the board, the writing team plays a significant role in deciding whether or not to accept certain feedback for revision.

The most important phase in the revision process is a yearlong pilot experiment in schools. The purpose of the experiment is to confirm whether the new series of textbooks would be appropriate, to collect data and feedbacks for revisions of textbooks and teacher guidebooks, and to have the experiment schools play a role in demonstrating the use of the new textbooks. The number of participating schools is about 30. These are either recommended by superintendents of local school districts and then designated by the minister, or they are elementary schools attached to national universities that specialize in teacher education. Given that the current mathematics textbooks were developed sequentially, the piloting in schools happened sequentially. For instance, in 2000, the teachers of third- and fourth-grade students were involved in demonstrating implementation of textbooks.

There are two periods for seminars and teacher training sessions related to the school experiments. Teachers are supposed to prepare two kinds of feedback, an overall evaluation of textbooks and teacher guidebooks, and then a description of specific problems and possible solutions with page numbers and lines of textbooks. The overall evaluation consists of five areas: appropriateness, correctness, connection, novelty, and others, as follows:

1. Appropriateness
   - Are contents appropriate to achieve the purposes of the national curriculum?
   - Are contents suitable for students' development, need, interest, and ability?
   - Is the amount of content balanced with the available teaching time?
   - Are the levels and difficulty of content appropriate?
   - Can instructional materials be easily used?

2. Correctness
   - Are illustrations, pictures, tables, and statistics selected correctly?
   - Are contents or terms correct?
   - Are contents correctly constructed from knowledge system of the subject?
   - Are contents unbiased or consistent?
   - Are there errors or typos?

3. Connection
   - Are connections across grades and within a subject reflected?
   - Are contents presented on the basis of interconnections among subjects?
   - Is the arrangement and organization of contents integrated?

4. Novelty
   - Are contents appropriate for students' independent learning?
   - Are the selection and presentation of contents novel and appropriate from the perspective of students?
   - Are contents realistic and creative?
   - Is the overall design original?

5. Others
    • Are there any kinds of prejudice in region, religion, or gender?
    • Are there contents that need to be added?
    • Are there contents that should be omitted?

Approval of Secondary Mathematics Textbooks

As secondary mathematics textbooks are developed under the approval system, it is the writing authors' role to revise their textbooks. The writing authors make every effort to develop a good textbook and the approval committee evaluates whether or not the textbook is appropriate to be used at secondary school levels. This is different from the process of developing elementary mathematics textbooks in which multiple revisions are made in conjunction with feedback both from the board and from the pilot implementation in schools. Table 7.4 shows the steps, contents, and conditions in the process of approval (MOE, 2000).

Note that the first main examination includes not only common criteria but also subject specific criteria in six areas with 20 evaluation perspectives. The followings are evaluation perspectives for secondary mathematics textbooks:

1. Does the textbook sufficiently reflect the nature, objectives, contents, teaching and learning methods, and assessment of the national curriculum?

2. Is the textbook organized in a way to consider students' individual differences so that students learn mathematics at their own pace and ability?

3. Does the textbook provide contents in a way for students to attain their learning goals?

4. Does the textbook consider mathematical connections so that stepwise or level-based learning is fostered effectively?

5. Are the contents adequate for the amount of learning in one class period?

6. Are there any errors of content or inaccurate theories?

7. Does the textbook provide students with reading materials that may help them recognize the usefulness of mathematics?

8. Does the textbook select contents related to real-life contexts and consider connections with other related subject areas?

9. As needed, does the textbook reflect learning areas applicable to all subject matters such as citizenship, humanity, environment,

Table 7.4.   Overall Process and Conditions of Approval

Steps	*Contents of Evaluations*	*Conditions*
Acceptance of draft to be approved	• Format and number of draft • Many related documents • Payment of a approval fee	→ Confirm whether any necessary paperwork is missing.
Basic examination	• Contents examination • Korean language notation • Vocabulary check • Analysis of editing	→ This examination serves as a bottom-line criterion for main examination.
Main examination (1st)	[Common criteria] • Constitution-observance • Educational law-observance • Copyright • Content-validity	→ Every evaluation item needs to be passed.
	[Subject-specific criteria] • Reflection of curriculum • Selection and organization of contents • Teaching and learning method • Representations and notations • Editing • Innovation	→ Innovative textbooks are recommended. → Individual evaluation of each item in each criterion in terms of A, B, or C. If a draft receives more than 2 Cs, it cannot be approved.
Main examination (2nd)	• Confirm whether feedback from the 1st examination are reflected	→ If a draft does not reflect feedback from the 1st evaluation, it is disapproved.
Teachers' guidebook evaluation	• Same as the process of textbooks	→ If a guidebook is disapproved, its original textbook is regarded as disapproved.
Final evaluation	• Final decision of textbooks and guidebooks—approval/ disapproval	→ Both textbooks and guidebooks need to be approved.
Announcement		→ Provide reasons for disapproval.

> economy, energy, consumer, career counseling, information ethics, and gender equity?

10. Does the textbook provide various and effective teaching and learning methods to foster logical thinking, inquiry, problem solving, creativity, reasoning, and applications based on mathematical concepts, principles, and rules?

11. Does the textbook properly present data for learning activities, and provide methods of data collection, analysis and application?

12. Does the textbook effectively use various manipulative materials and instructional technology including calculators, computers, and the internet?

13. Does the textbook introduce assessment methods and tasks that are compatible with the objectives, contents, and methods of mathematics education?

14. Does the textbook abide by the orthographic rules of Korean and foreign languages as well as the norms of standard language?

15. Does the textbook correctly use mathematical terms and notations as presented in the national curriculum without omission or repetition?

16. Is the textbook edited innovatively and does it use space effectively?

17. Are the photos and illustrations clear and compatible with the contents?

18. Does the appearance of the textbook including format, length, and chromaticity adhere to the guidelines?

19. Does the textbook take advantage of novel ideas and organize contents in an innovative way?

20. Are teaching and learning processes as well as assessment methods creative?

With regard to the process of approval, two aspects need to be emphasized. First, approval criteria include an area of innovation in order to encourage textbook writers to pursue novel ideas contrary to long-term practices of textbook development. This emphasis extends to editing and appearance. Previous textbook development had focused mainly on the contents per se, current development regards not only contents but also overall design.

Second, the approval of teacher guidebooks is necessary. Given that guidebooks are the most important instructional resources to teachers, the approval committee mandates a thorough evaluation of guidebooks. This prevents publishing companies from producing low-quality guidebooks while hurrying to develop students' self-learning books for commercial purposes. In fact, there are 20 detailed evaluation perspectives related to the introduction of the curriculum and the textbooks, format and organization, teaching and learning methods, usage and guide of data, representation and notation, editing, and innovation.

Alignment With the Curriculum

As indicated above, the review or approval process is designed to insure that the textbooks are aligned with the national mathematics curriculum framework. In brief, the review and assessment board works for elementary mathematics textbooks and the approval committee works for secondary mathematics ones. Since there is only one series of elementary mathematics textbook developed by one institute, the textbook is aligned with the mathematics curriculum framework through the efforts of the writing team under the board's supervision.

In the case of secondary mathematics, if textbooks are not aligned with the mathematics curriculum framework, they cannot be approved and thus cannot be used in schools. In order to get approved, the textbooks should not receive more than 2 Cs with regard to the 20 evaluation perspectives described above. Since the perspectives from 1 to 13 have a direct or indirect relation with the curriculum framework, weak alignment with the curriculum results in the textbook not being approved. Authors who fail an initial evaluation may resubmit for approval.

Until the mid 1990s, the ministry limited the number of approved secondary textbooks. As a result, for example, only 8 out of 43 mathematics textbooks submitted for use in middle schools were approved. Currently, the approval committee of each subject determines the number of textbooks to be approved. The result is that textbooks are approved as long as they meet the 20 criteria (Ministry of Education and Human Resources Development, 2000). The overall approximate approval rate is 1 out of 3 for middle school mathematics textbooks, and 1 out of 2 for high school textbook. This low rate is mainly due to the fact that the approval committee evaluates the 20 criteria in a strict and rigid way to select textbooks of good quality.

TEXTBOOK SELECTION AND USE

Selection and Payment

Because there is only one elementary mathematics textbook series, every elementary school uses the same textbooks. However, in the secondary case, fairly selecting one out of the many authorized mathematics textbooks is of great concern. Usually a committee comprised of all mathematics teachers in a school decides on the one textbook to be used.

The exact selection process varies from simple discussion to complex but systematic procedures. For example, each committee member may look over all mathematics textbooks approved by the government and score them individually. The textbook receiving the highest average score is finally chosen. Alternatively, each member may score the approved textbooks, and then the committee provides strengths and weaknesses of the top three textbooks to the committee of school management, which finally selects one of the three.

The government pays for mathematics textbooks for elementary and middle school students, whereas parents pay for textbooks for high school students. If someone needs mathematics textbooks, she or he can buy them at a large bookstore at a very cheap price. Basically, students own their mathematics textbooks at all school levels. Recently, used textbooks tend to be collected at the end of the year not for use next year but for recycling. As some students want to keep the textbooks in which they have made personal marks in the process of learning, the collection of textbooks is partial and voluntary.

Use of Textbooks

As described, Korean textbooks play a significant role in determining what teachers teach and students experience. To emphasize, our textbooks may be regarded as a bible that shapes the process of teaching and learning. Textbooks have an absolute impact on students' mathematical experience. Teachers are expected to cover all the mathematics topics presented in each grade level textbook.

Teacher Orientation to New Textbook

After the new curriculum is ready for national use, professional development is provided to help teachers become familiar with the curriculum and use it effectively. Professional development is implemented in a rather top-down format. Selected mathematics teachers are informed of the changes in curricular emphases and the subsequent instructional implications, and then these teachers inform their colleagues. As the teacher-training program is general and at best spends only a short time dealing with each subject matter, it is not likely that teachers will develop deep understanding of the curriculum in this first exposure and be ready to fully implement it. It demands lots of commitment and effort on the part of the teacher.

The professional development emphasizes the teacher's autonomy in implementing the curriculum, pointing out that curriculum materials including textbooks are resources that teachers use to achieve their personal instructional goals and priorities. In other words, teachers are expected to improvise in teaching as long as they address the required content of the curriculum for their grade. In fact, many levels of reconstruction are encouraged with regard to the implementation of the curriculum. The national mathematics curriculum is interpreted locally by provincial educational offices, and in turn by each individual school, and finally reconstructed by the teacher.

Use of Textbooks by Teacher and Students

Teachers use the textbook as the most important teaching material. Although they are encouraged to reconstruct the textbook, teachers tend to teach what the textbook says. The reconstruction of a textbook usually is limited to substituting manipulative materials for what the textbook introduces, changing the sequence of several activities within a class period, or adding/deleting some units or topics.

Students also regard the textbook as the main material to learn mathematics. In particular, elementary school students tend to solve every problem both in the textbook and in the workbook (although some series of problems are designed only for low-achieving or for high-achieving students). Secondary school students use the textbook to learn basic concepts and skills in the first place, and then tackle self-study or the examinee's book published by the company who supplies the textbook.

It depends on the teacher whether she or he implements the national curriculum and the textbooks as intended by the writers and as expected by the government. Since the recent national curriculum was implemented, KICE has been conducting multiple studies on the curriculum implementation at the elementary, middle, and high school levels through the review of curriculum materials produced by local educational offices or individual schools, questionnaires for teachers and supervisors, interviews with teachers, and classroom observations. Such studies examine the degree to which the particular characteristics of the curriculum (e.g., student-centered, reconstruction of textbooks) are being implemented as intended, and probe what are the main obstacles in the process of implementation, which may point out general directions to be considered in developing the next curriculum.

The more teachers understand the intention of the new curriculum, the greater the likelihood they will implement it as intended. On the one hand, there has been a report that teachers generally are aware of the overall characteristics of the curriculum and attempt to activate the ideas of learner-centered teaching in their mathematics classroom as reflected in teaching objectives, contents, instructional methods, and assessment (Sung et al., 2003). On the other hand, institutionalization of the new curriculum ideas at the local school level has been insufficient, for instance as evidenced by teachers' limited reconstruction of the contents to be taught (Pang, 2005). Continuous efforts are needed to get teachers involved actively in the full implementation of the curriculum.

CLOSING REMARKS

Due to the importance of mathematics textbooks in the Korean context, we have consistently made great efforts to develop good ones. This section deals with current concerns and future directions with regard to the development and implementation of instructional resources.

Contents and Design

Good quality of textbooks encompasses both contents and design. With regard to content, textbooks emphasize or reflect what the national curriculum intends to do. In the case of elementary mathematics textbooks, alignment with the curriculum is checked and confirmed by multiple means such as discussion and seminars within the writing team and the research team, review by the national review board, school based implementation, and so forth. In the case of secondary mathematics textbooks, alignment is one of the most important criteria used in the process of the government approval procedure.

A concern for design is relatively new in comparison with that of content. On the excuse that they are cheap or free, the government has restricted textbooks to low quality paper, a limited number of color, and economical size. However, current textbooks attempt to induce students' engagement and interest by considering editing and design of textbooks, in keeping with various general books. In fact, current textbooks have upgraded the quality of paper, permitted the use of various colors, and changed the size in a way for students to use easily and effectively. Professional people have called for better editing of textbooks.

Teacher Involvement

Teacher involvement in textbook development has been called for and has indeed increased. As was noted earlier, more than 50 teachers from elementary schools were involved in writing the current mathematics textbooks. Almost all teams developing secondary mathematics textbooks include middle or high school teachers. In addition, approximately half of the approval committee is secondary school teachers.

However, despite the number of textbook authors who are secondary school teachers, the breadth of real engagement is rather limited in that university professors determine and control the overall directions of development. Teachers usually write specific chapters within the guideline. Although a horizontal collaboration between professors and teachers is encouraged for the integration of theory and practice, a vertical collaboration seems dominant, which may not always be the intention of the professors.

Another issue is the workload of textbook authors who are teachers. As teachers do not have released time from their regular jobs to author textbooks, they may find it difficult to focus on textbook development, regardless of their commitment and values. Greater institutional efforts are needed to get teachers as partners with researchers to fully engage in authoring textbooks. In the case of elementary mathematics textbooks, it may be useful for the teacher writers to be sent to the research and development institute that is in charge of textbook development so they can concentrate on authoring textbooks without worrying about their daily teaching.

Regular Research and Textbook Revision System

Although there has been research on textbook design and implementation, its regularity and depth have been insufficient. Basic research is needed concerning what are the essential contents in mathematics to be taught regardless of the change of the curriculum, what are the critical criteria in the process of textbook approval, what are the main functions and types of textbooks, and what are the models of textbooks reflecting the current trends in mathematics education.

More comprehensive study is called for with regard to the various needs of society, the overall educational environment, students' learning of mathematics at each grade level, and others. More systematic and regular surveys and interviews are necessary with regard to textbook usage and the concomitant analysis of strengths and weaknesses of the textbook. The lack of connection between research and textbook development has been really problematic, so all studies should be directly connected to subsequent textbook development.

Another concern is the development of a regular textbook revision system. Textbooks change whenever the curriculum changes. In other words, once developed or approved, the textbook is used until the new curriculum evolves. For this reason, the field test of textbooks needs to receive greater emphasis and scrutiny before developing the final version of the textbook. Beyond mere teacher interviews and surveys, classroom observations and subsequent analyses should be made in detail. Such methods can urge both the writing team and the review team of textbooks to be sensitive about what works in the actual classroom situation and what does not, so that they develop better textbooks grounded in classroom contexts.

In addition, a constant evaluation or revision system needs to be implemented while the textbook is being used. Such a system should deal with not only the errors or mistakes of textbooks but also the degree of fit with the curriculum, the opportunity to learn provided by the textbook on the basis of students' mathematical abilities, the number of main learning themes, the appropriateness of descriptions and representations, the style of textbook editing, and so. For these purposes, an online system has recently been made by which anyone can report errors, mistakes, or suggestions concerning textbooks. Although textbook development per se has been the sole focus, textbook revision is now being called for.

Variety of Textbooks Developed

"Variety" may imply three different meanings. First, it means that there are various textbooks so that schools or teachers can choose. Second, it means that there are various constructions of contents so that students can learn mathematics by their ability or learning theme. Third, it means that there are various media incorporated so that students can learn mathematics not only with traditional textbooks but also with multimedia.

Government Issued Versus Government Approved

The first meaning of variety is directly related to the issue of government issued or government approved (or even privately issued) textbooks. In the case of elementary mathematics textbooks, it is an open question as to whether multiple teams authoring a variety of textbooks (as for secondary textbooks) would be better than the current system. On the one hand, this kind of development method might lead to diverse and creative textbooks, partly because free competition among many experts is allowed. On the other hand, this method makes it difficult to control the contents to be taught and their levels in a systematic and consistent way, and it requires more financial support.

It may be a practical issue in that there are lesser funds for research and development of textbooks and fewer researchers in comparison with international contexts. Given this, we continue to develop one elementary mathematics textbook series, while attempting to maximize the quality of the product thorough planning, conducting research before development, assessing multiple textbook development plans, monitoring the process of textbook development, managing scores of school-based experiments, involving many teachers and mathematics educators, and others as mentioned earlier.

In the case of secondary mathematics textbooks, the main issues are the criteria and the process of approval. First of all, the criteria should be comprehensive and strict enough to make sure that textbooks are aligned with the national curriculum. At the same time, they should be inclusive enough to encourage autonomy and creativity on the part of textbook authors. As described, recent criteria include creativity, but the actual products are less creative than what has been expected. Probably this is because textbook authors were unwilling to risk major changes from the previously approved textbooks, being afraid that some conservative committee members may reject the unusual. Consequently, at one level, the approved textbooks are innovative in their use of different episodes or illustrations to introduce a mathematical concept and in their more polished and learner-friendly formats. At another level, the textbooks are the same in that the main mathematical theme is taught and practiced in a similar way. Given this, we ponder how to elicit textbook authors' creativity and authenticity within the government-approval system.

The process of approval should be fair and clear. One problem we face is staffing an approval committee because many professors and teachers are engaged in textbook authoring (K. H. Lee, 2004). Another problem is that the limited number of committee members usually do not have enough time to carefully examine the textbook to be approved. To address these problems, we attempt to recruit more teachers for these committees and to divide the approval procedure into preliminary, basic, and main examination, thereby allotting different experts to different stages of examination.

Textbooks Tailored to Students' Ability

The current curriculum promotes student learning of mathematics appropriate to their mathematical ability. However, a single textbook covers both basic mathematics to be taught for all students and applied mathematics for high-achieving students. This makes students develop unreasonable beliefs that they should cover all the contents in the textbook. These low-achieving students are attempting to deal with a deeper level of mathematics. The curriculum is built on a level-based differentiated structure in which students move to the next level only after they have mastered the current level, the textbook is not appropriate for students who do not pass the current level. Given this, different schools and teachers develop their own instructional materials for low-achieving students, which turns out to be extra work for teachers.

Against this background, we move on to the discuss whether we should develop three different mathematics textbooks and accompanying learning materials tailored to different mathematical ability levels. The issue here is how to evaluate students who have studied mathematics with different textbooks. If we measure only the basic and fundamental mathematics, both low-achieving and high-achieving students will lose their taste for mathematics that is appropriate to their ability. Similarly, if we evaluate in a way that is sensitive to students' ability, we face the problem of how to decide relative rankings among all students, which is one of the main resources for college entrance.

Different Types of Textbooks

The current curriculum emphasizes that the textbook is to be the main instructional resource, but still one among several resources. Schools or teachers are indeed to be encouraged to change certain aspects of the textbook as long as it is in line with the basic principles of the curriculum. However, teachers usually depend exclusively on the textbook. This encourages us to develop different textbooks and other materials so that teachers can select according to their own priorities and instructional goals. Toward this end, e-textbooks and related materials are being developed.

Appropriate usage of information and communication technology has been called for but actual implementation is rather weak, partly because there are not enough instructional materials that may be directly connected to daily mathematics lessons, and because teachers do not fully understand how to integrate such technology into their instruction. Different types of textbooks and activity files using multimedia are being created, instead of relying on vague expectation that teachers will construct such materials for themselves.

Students' Interest and Dispositions

Another important issue to consider in the textbook development is how to promote positive student mathematical disposition. Korean students, even with excellent mathematics achievement, develop increasingly negative mathematical dispositions and feel a lack of self-esteem with regard to their mathematical ability. In fact, the TIMSS and TIMSS-R found that the countries with the highest performance in mathematics, including Korea, also had students with the most negative perceptions of mathematics and success in the subject (Mullis et al., 2000).

As a part of attempts to elicit students' participation of and interest in mathematics, current textbooks include more concrete activities, various games and puzzles, episodes related to students' daily lives, and historical content related to the learning theme than did the previous textbooks. However, many teachers think that it is not enough. It is time for textbook-writers to ponder carefully whether the activities of textbooks are appropriate to provoke students' interest, whether they are easy to implement in any classroom situation, whether they can be accomplished within a classroom session, and so forth.

Teacher Guidebooks

Teacher guidebooks are the main means of determining the degree to which textbooks are used in a way the authors intended. They are the foremost resources for a teacher to understand background information, to check the main objectives, to design an instructional flow, and to assess students' attainment. In this sense, guidebooks are more powerful than textbooks in shaping the process of teaching and learning.

However, surveys consistently show that teachers are less satisfied with guidebooks than textbooks (Sung et al., 2003). Whereas guidebooks are clear in describing learning objectives and contents to be taught, they are weak in introducing various instructional materials and references, illustrating when and how to use such materials to maximize the learning goals, explaining effective teaching and learning methods specific to a given mathematics content, and guiding how to evaluate students' performance. Given that the teacher is the person who makes the curriculum come alive in the classroom, the guidebooks should include more detailed explanations of curriculum and various instructional materials which teachers are ready to implement in the classroom. Inherent in this is that guidebooks should include the best information and richest materials, including different activities for a wide variety of situations, detailed illustrations of how to implement them, and articulated guidelines of how to assess students. This does not mean to underestimate a teacher's professionalism to orchestrate his or her instruction. Rather, with rich materials the teacher can use the guidebooks in ways that are consistent with his or her particular teaching style, the particular group of students, and the constraints and supports of particular school contexts.

In fact, we focus more on textbook development than on teacher guidebook. Especially for secondary mathematics textbooks, once the textbook gets to be approved, its concomitant guidebook also is easily approved. In addition, textbooks go through a systematic revision procedure and are open to criticism even after they are developed or approved, whereas guidebooks have little chance for such revision or evaluation, mainly because individual teachers use them optionally. Given the importance of teacher guidebooks in shaping mathematics instruction in the classroom, the urgent need is how to develop a good quality of guidebook as well as that of textbook.

REFERENCES

Bae, J. S. (2004). *A clown's performance*. Paper presented at the 10th International Congress on Mathematical Education, Copenhagen.

Grow-Maienza, J., Beal, S., & Randolph, T. (2003, April). *Conceptualization of the constructs in Korean primary mathematics*. Paper presented at the annual meeting of American Educational Research Association, Chicago.

Kim, Y. M. (1999). Comparison and analysis of mathematics curriculum for lower graders (in Korean). *Journal of Educational Research in Mathematics, 9*(1), 121–132.

Lee, K. H. (2004). Development and characteristics of Korean secondary mathematics textbooks. In Korea Sub-commission of ICMI (Ed.), *The report on mathematics education in Korea* (pp. 69–84). The Korea presentation at ICME-10. Copenhagen.

Lew, H. C. (1999). New goals and directions for mathematics education in Korea. In C. Holes, C. Morgan, & G. Woodlouse (Eds.), *Rethinking the mathematics curriculum* (pp. 218–227). Philadelphia: Falmer Press.

Ministry of Education. (1998). *Commentary on elementary mathematics curriculum* (in Korean). Seoul, Korea: Dean Textbooks.

Ministry of Education and Human Resources Development. (2000). *A textbook white paper* (in Korean). Seoul, Korea: Dean Textbooks.

Ministry of Education and Human Resources Development. (2001). 3-*Ga mathematics textbook* (in Korean). Seoul, Korea: Dean Textbooks.

Mullis, I. V. S., Martin, M. O., Beaten, A. E., Gonzales, E. J., Kelly, D. L., & Smith, T. (2000). *TIMSS 1999 international mathematics report*. Boston: Boston College.

Pang, J. S. (2004). Development and characteristics of Korean elementary mathematics textbooks. In Korea Sub-commission of ICMI (Ed.), *The report on mathematics education in Korea* (pp. 37–67). The Korea presentation at ICME-10. Copenhagen.

Pang, J. S. (2005). Transforming Korean elementary mathematics classrooms to student-centered instruction. In H. L. Chick & J. L. Vincent (Eds.), *Proceedings of the 29th conference of the International Group for the Psychology of Mathematics Education* (Vol. 4, pp. 41-48). Melbourne, Australia: Psychology of Mathematics Education.

Park, K. M., & Yim, J. H. (2002). Comparative study of the mathematics textbooks of Korea, Japan, the United States, and England (in Korean). *School Mathematics, 4*(2), 317–331.

Sung, K. H., Jeong, K. H., King, D. H., Chow, S. H., Quick, Y. S., & Chow, J. H. (2003). *A study on the seventh national curriculum implementation at the elementary school level* (in Korean). Seoul: Korea Institute of Curriculum and Evaluation.

Woo, J. H., Lew, H. C., Moon, G. H., & Park, K. M. (2002). *Teacher guidebook of high school mathematics (10-Na)*. Seoul, Korea: Daehan Textbooks.

Yim, J. H., Lee, D. H, Lee, Y. R., Park, S. K., & Jeong, Y. K. (2004). *An analysis and evaluation of the content relevance in the primary and secondary school mathematics* (in Korean). Seoul: Korea Institute of Curriculum and Evaluation.

CHAPTER 8

CURRICULUM DEVELOPMENT IN CHINA

Perspectives From Curriculum Design and Implementation

Jun Li
East China Normal University

Curriculum has several dimensions (Goodlad, Klein, & Ty, 1979; Robitaille et al., 1993; Valverde, Bianchi, Wolfe, Schmit, & Houang, 2002). Curriculum could be a set of documents, such as syllabi or standards, highlighting official intentions, aims, and goals. It could be scholars' ideal educational designs based on their research or new ideas. It could be teaching strategies, practice, and activities that actually go on in school and in the classroom. It could be considered as student achievement, assessed after learning. Finally, curriculum could be considered as the textbooks and other organized resource materials used by schoolteachers and students. This paper is mainly concerned with textbooks as potentially implemented curriculum.

Mathematics Curriculum in Pacific Rim Coutries—China, Japan, Korea, and Singapore: Proceedings of a Conference, pp. 127–140
Copyright © 2008 by Information Age Publishing
                    127

A VERY BRIEF HISTORY REVIEW

Mathematics education in China has a long history that goes back to ancient times. As early as the *XiZhou* dynasty (more than 2,000 years ago), "six arts" (rites, music, archery, chariot-riding, calligraphy, and arithmetic) education formally existed in China. It should be noted that here "arithmetic" not only referred to arithmetic per se, but also included geometry and some other scientific knowledge such as astronomy and the calendar (Ma, 2001). Similar to the status of arithmetic and geometry in Western "seven liberal arts," the status of arithmetic in the six arts was lower. The scientific values of mathematics education were undervalued in ancient times.

Comparable in significance to Euclid's *Elements* in the West, the classic work "The Nine Chapters on the Mathematical Art" exerted the greatest influence on the science of mathematics and its education in China. It established the traditional mathematics style that was very useful in application and calculation. The Nine Chapters contains 246 questions about land area, exchange rates between goods, distribution by ratio, square roots and cube roots, volume of solids, profit and loss, linear equations, and right triangles. The texts are firmly based on practical needs at that time.

The Nine Chapters was one of the most popular standard mathematics textbooks in China until European mathematics was introduced in about 1600. In 1607, the first six volumes of Euclid's *Elements* were translated into Chinese by the Chinese mathematician Xu Guangqi and the Italian Jesuit Priest Matteo Ricci. After that, especially after the Opium War in 1840, Western mathematics and modern school systems were introduced. In 1862, the first modern school in China was run by the government.

Until 1906, algebra textbooks in China still used the traditional page layout in which the text reads vertically from top to bottom and horizontally from right to left. The constants and variables were denoted by Chinese characters instead of Roman letters. After the 1911 Revolution, elementary mathematics was offered in almost all schools (Mo & Xu, 1987). In the modern education systems, traditional Chinese mathematics textbooks were replaced gradually by Chinese translations or adaptations of European, Japanese and American textbooks. For instance, during the 1930s and 1940s, *College Algebra* written by Fine and *Plane Geometry* and *Solid Geometry* written by Schultze, Sevenoak and Schuyler were the popular textbooks in China. Also, *Plane Trigonometry* written by Granville and *New Analytical Geometry* written by Smith, Gale and Neelley, were also widely adopted (Mo & Xu, 1987; Ngai, Lee, & Ko, 1987).

In 1949, the People's Republic of China was established. Because of political reasons, the Soviet Union's model of education was imported. Almost all of the school textbooks used at that time were adapted from the Soviet Union's school textbooks by the People's Education Press (PEP) which is directly affiliated with the Ministry of Education (MOE) in China. A national unified textbook policy became effective in 1952. The PEP served as the only official developer of national curriculum and teaching materials until the late 1980s (Li, 2004a).

Since the Soviet Union 10-year educational system was not in accord with the 6-3-3 system in China, the imitation of the Soviet curriculum resulted in the reduction of teaching content. After 1958, China's curriculum developers tried to redesign school mathematics based on the practical conditions in China. They attempted to keep a balance between Confucian and Western-style education. However, the following characteristics of the Soviet Union's curriculum system framework were widely accepted, insisted, and even developed by Chinese educators: integrity, coherence, focus and rigor. The syllabus issued in 1963 and textbooks published during 1958–1963 reflected these characteristics. Indeed, the influence of the Soviet Union remains great in China.

The worldwide "new math" movement did not really occur in China during the 1960s and 1970s. Prior to 2001, algebra and geometry were the two main teaching contents at school level. They were taught simultaneously with separate textbooks and sometimes even by different teachers. The algebra textbooks emphasized the rules and regulations of basic knowledge. The geometry textbooks reflected a typical "theorem-based curriculum," and the geometric reasoning is typical of "theorem-based deduction" (Bao, 2004). All Chinese students had about 10 teaching hours in statistics education in Grade 9, but most of them did not learn probability at the secondary school level before 2000.

In the new standards-based textbooks, teaching time for statistics and probability was increased significantly. Using our textbooks (Wang, 2001) as an example, the teaching time allotted to statistics and probability is increased to 69 total teaching hours at the junior high school level (Grades 7-9). For the six semesters, the teaching hours allotted to introduction, algebra, geometry, project learning, and stage review were 4, 156, 133, 24 and 24, respectively, with each teaching hour lasting 45 minutes. The arrangement for the other sets of textbooks might be different but should be roughly the same. At the senior high school level, according to the new standards (Ministry of Education of China, 2003), about 24 teaching hours are devoted in a required mathematics course to the study of statistics and probability.

Overall, the Chinese education system is a national system governed by the Ministry of Education. Before 1986, only one series of mathematics textbooks, compiled and published by the PEP were used at the same time in all schools in mainland China. However, central control was made less rigid in the late 1980s. For instance, Shanghai was allowed to have its own local official curriculum and to organize separate university entrance examinations for Shanghai senior high school students. But major changes happened after 2000. For mathematics teaching, we now have six series of national standards-based textbooks being used in primary schools. Another nine series and six series of national standards-based textbooks are being used in junior high schools and senior high schools, respectively.

THE DEVELOPMENT OF A SERIES OF TEXTBOOKS

Since the MOE introduced competition in providing school textbooks, the PEP and some provincial publishing houses, as well as some major normal university publishing houses have been very active in developing new textbooks and other teaching and learning materials. Once a publishing house has decided to create a new series of school textbook and has organized a writing team for it, a proposal and sample chapters of the textbook are required to be submitted to the MOE. If the project is approved, the textbook writing and a small-scale experiment can start. Then the publishing house is usually required to provide a report indicating the results of the textbook experiment. Experiment reports have not been required since 2000, but criticisms for this have been raised recently. All textbooks have to be reviewed by National Teaching Material Authorization Committee. The committee will give detailed feedback to the publishing house. After modification, the textbooks have to be reexamined by the committee. Finally, if approved, the textbook (with its price) will appear on the MOE Web site for textbook selection nationwide.

Before 2000, the PEP textbooks were mainly designed and written by in-house editors. But after 2000, it seems that more textbook writers come from universities. Take the textbooks (Wang, 2001) I am working on as an example. We have a textbook writing team with almost 20 members, including university faculty members, teaching supervisors, schoolteachers, and editors. The general editor of the textbook is the president of my university. He is also a famous mathematician in China. Two subeditors and most of the writers are university faculty. The workload for textbooks authors is heavier. In addition to authoring the textbooks, they have to do their regular jobs and give talks to schoolteachers during school break periods.

There are two semesters per school year and we use one textbook in each semester. That means we prepared six textbooks in all for junior high school students enrolled in Grades 7 to 9. More than 10 people worked on each book. Since students bought their own textbooks before each semester, it was feasible for us to make careful revisions every semester. Therefore, the newest version of a textbook was always readily used. The changes between the different versions were minimal but necessary and included changes such as data updating or language editing. In addition, we also wrote teacher's manuals and student exercise books. All answers or explanations to the questions, exercises, and problems that appeared in our textbooks were provided in the teacher's manuals. Some schoolteachers who had used the textbooks were also invited to cooperate with us in writing the teacher's manuals and hypertexts in electronic devices, such as CD-ROMs. Other supplemental materials are also available from teacher journals, problem booklets, workbooks and the Internet. Well-prepared teaching plans, excellent examples, and well-designed workbooks were especially welcomed by schoolteachers.

TEXTBOOK DESIGN AFTER 2000

Curriculum development after 2000 in China has been mainly driven by international comparative studies and guided by university professors. The syllabus/standards issued in other countries or areas such as Australia, Britain, France, Germany, Hong Kong, Japan, Korea, Russia, Singapore, Sweden, Taiwan, and the United States were studied carefully. Some publications such as *Mathematics as an Educational Task* (Freudenthal, 1973), *Mathematics Counts* (Cockcroft, 1982), *Everybody Counts* (National Research Council, 1989), and some textbooks such as those by the University of Chicago School Mathematics Project (1992) were translated or introduced to China. We also learned a lot from other countries through international exchanges. We have to say that the following changes in our curriculum are the result of learning from other countries: to introduce a credit hour system at the senior high school level, to include statistics and probability, calculus, and algorithms into the school mathematics curriculum, to connect mathematics with daily life situations, as well as others.

The current curriculum reform was under the guide of two national curriculum standards issued by the MOE. One is *The Standards of Mathematics Curriculum for Compulsory Education* (MOE, 2001) and the other is *The Standards of Mathematics Curriculum for Full-time Senior High Schools* (MOE, 2003). Textbook writers made many special efforts to follow the reform principles indicated in the two standards. These include emphasizing "big ideas," designing some key concepts designed in a spiral way, connecting mathematics to both students' mathematical reality and their living reality, trying to make learning mathematics attractive, utilizing new technology in teaching and learning, and cultivating students' positive emotion and attitudes. I would like to illustrate this further by some examples selected from the chapters of statistics and probability.

Before the reform, the position of teaching probability in China was low and the teaching approach was theoretical and oriented toward numerical calculations involving permutations and combinations. The "frequency of occurrence," empirically derived definition of probability was despised, since it was thought of as posterior and inexact. Chinese students almost had no experiences of being involved in any concrete activities and simulations. In a study, it was found that some Chinese students did not believe chance can be quantified, even though they had learned some techniques of calculating probability (Li, 2000). To change this, the new national standards strongly recommended that statistics education should involve analysis of genuine data (MOE, 2001, 2003). Without the mention of variance, the teaching of probability is almost the same as the teaching of algebra.

In our textbooks, some "big ideas" were emphasized repeatedly, such as reasoning based on data and estimating probability from the relative frequency of occurrences of an event in a large number of trials. Here are four teaching examples or activities selected from different books that are used to improve students' understanding of probability gradually.

Example 1, Picture Match Education

Here are three pictures of equal size. Cut each of them equally into two pieces. Put the six pieces into a bag and mix them thoroughly. Pick out two pieces without looking. Do you think it is very likely, unlikely, or equally likely for you to get a pair that makes up one of the original pictures? Make a guess. How many trials on average do you need to get a match?

This activity, modified from a problem in *Teaching Mathematics* (Sobel & Maletsky, 1991), was arranged at Grade 7. This was the first year for Grade 7 students involved in the new curriculum. They had never been required to give their personal evaluations of the likelihood of chance phenomena but they were happy to do this and looked forward to the experiment results. In this stage, the students were required to or expected to

- record their own data on the blackboard and make their judgments based on the whole data set;
- find that the number of trials needed to get a match is varied but not totally random;

- compare the results of actual trials to their own guess and adjust their intuitions according to the results; and
- compare relative frequency to 50% to describe chance qualitatively: very likely, unlikely, and equally likely.

Example 2, Different Size Spinners

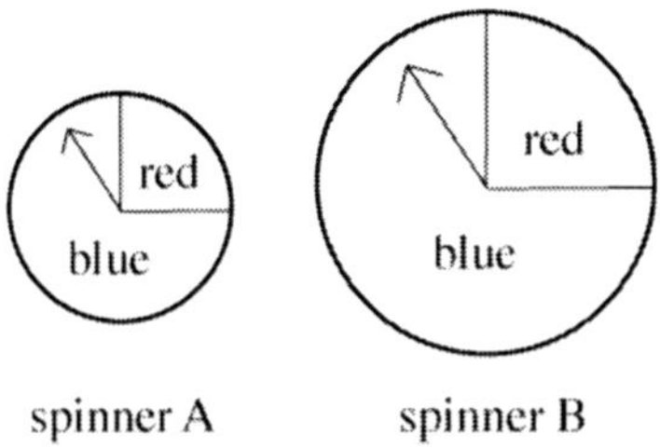

Spin each spinner's arrowhead as hard as you can. Suppose you want the arrowhead to stop in the blue part. Which spinner will give you a better chance for success?

Consider the following two responses given by students:

- Spinner B gives a greater chance because the spinner is bigger and has more blue part.
- The two spinners give the same chance because there are only two possible outcomes: land on red or blue. Each spinner has 50% chance to be successful. So it doesn't matter whether you choose the big one or the small one.

This problem was arranged at Grade 7, too. Let me address the second part of the problem first. In China, some master teachers often make use of students' errors and misconceptions in their teaching. They present some wrong answers, then encourage students to examine and discuss them carefully. Many Chinese educators believe this is helpful to improve students' understanding. However, schoolteachers were unclear about students' main misconceptions of probability since this was their first teaching of the topic. Therefore, in our textbooks, we cited some students' incorrect responses observed in our own research, as well as those reported in the literature. Teachers found this information about misconceptions is useful for their teaching.

To solve this problem, students are required to

- think about which spinner's arrow has a greater chance of landing on the blue section;
- make a judgment about the given student responses according to their own experiences;
- perform the experiment in pairs: each student was required to spin 25 times with each spinner while their partner recorded the number of successes;
- pool data and make a cumulative relative frequency graph with two different colored pens to illustrate the probabilities for each spinner;
- examine the given student responses again, mindful of the new data; and
- report their own findings obtained from the activity and give their comments on the two responses cited above with explanations.

In China, the class size is often very big, normally about 60. Twenty-five spins for each student, totaling around 1,500 spins for the whole class, makes it feasible to get a good estimate of the expected probability when all the students' data are pooled.

Finally, another spinner with eight equally divided sectors (four red sectors and four blue sectors) was shown to students and they were asked whether they could predict the chance the spinner would land on one of the sectors colored red without performing the experiment. At the time of the activity, students had not yet been taught to calculate probability theoretically. However, when students reflected on the conclusion of the previous problem, realizing that both the spinners yield a 25% chance of successful, they solve the new spinner problem easily. We believe it is an important approach to develop students' understanding by posing some challenging tasks to them. Many such questions can be found in our textbooks' margins.

Example 3, Rock, Scissors, Paper

In China the game is called "Rock, Scissors, Cloth." The well-known game of Rock, Scissors, Cloth includes a circular hierarchy of winning plays. The rock crushes scissors, scissors cut cloth, and cloth covers the rock. If the same item is formed, it's a tie and the game is repeated.

We ask: Suppose each player plays each strategy with equal probability. What is the probability of getting a tie for each game? Try to use a tree diagram to solve this problem first. Then, apply the resampling method to estimate how often a tie will occur on average. Compare the two results to see whether they verify each other.

This problem was arranged in Grade 8 to follow the teaching of tree diagrams and interpretation of probability. The students were expected to

- conduct the analysis of the game by tree diagram;
- count the total number of equally likely alternatives and the number of alternatives to getting a tie;
- predict the probability of getting a tie theoretically;
- simulate the game by taking out chips from bags; and
- pool data and make a cumulative relative frequency table to estimate the probability of getting a tie.

After students had had experience with variations and realized that they could find trends across sampling, how to predict probabilities without experiments were introduced. In this stage, as shown above, the students are expected to solve the same problem with two different methods. According to our study (Li, 2004b), each of the two approaches has its limitations. When probability was taught in a theoretical and formal way, students' intuitions could not be modified directly. When probability was taught in an empirical way, students' knowledge of classical probability was poor and was not improved simultaneously. Therefore, it is practical to teach both theoretic and empirical definitions of probability at school level and necessary to make the connections between them.

Example 4, Win a Prize

A TV advertisement for ice cream says a promotion is running. They have sealed one of four pictures on each ice cream stick. Putting four different sticks together will show a bigger picture that you can exchange in some designated stores for a prize. Suppose the same number of each of the four types of sticks is made and that the sticks are chosen randomly in producing the ice cream bars. On average how many ice cream bars would you need to buy before you win a prize?

This problem is selected from a Grade 9 textbook. Students are expected to solve it by simulation. Let "1," "2," "3," and "4" stand for the four types of stick. Use a calculator to generate random whole numbers between 1 and 4 for simulations. Record each number generated, until all the four numbers appear, which means you win a prize. Stop and count how many ice cream bars you bought for the prize. Each student does 10 rounds of simulation and works out on average how many ice cream bars have to be bought to win the prize.

According to the new standards, all junior high school students should be able to use scientific calculators to process data. Calculators capable of generating random whole numbers are currently available in the market. So we try to encourage the students in the last year of junior high school to explore probability problems by calculator simulations, although this was not required by the national standards. The standards set the basic requirements. For the topics not mentioned in the standards, textbook writers and schoolteachers are allowed to include such topics as advanced learning material or supplementary learning material.

In our textbook, we encourage students to use calculators or computers to perform calculations, simulations, and make statistical graphs. In addition, detailed information about how to make statistical graphs and calculate measures of center or spread with Excel are presented as reading materials in our textbooks. The publishing house also provides electronic teaching materials for teachers and students. After engaging in actual experiments, teachers can run computer simulations on big screens to let the students see what will happen in the long run.

A finding reported by Bao (2004) is that the new textbooks pay more attention to student investigations and present knowledge in various contexts. In addition, he found the level of difficulty for "symbolic computation," "complex reasoning," and "topic coverage" has been reduced in the new junior high school textbooks. Limited by the length of this paper, no further explanations are included here.

INVESTIGATIONS IN TEXTBOOK DEVELOPMENT

China is still at the very beginning of forming its new curriculum. Before 2000, the PEP series dominated the textbook market, constituting more than 70% or more of the market's supply. Now, there are textbook selection policies. Usually, all schools within a city select the same textbook series in the same year. A textbook selection committee always is chaired by a local educational bureau director and composed by several members who are teaching supervisors, headmasters, classroom teachers and parents. They come and sit together, read the textbook evaluation report provided by the province, read and compare textbooks by themselves, discuss and vote, and make their final decision in 2 days. In China, students buy their textbooks, exercise books, and calculators. The process of textbook selection seeks books that have a low price, have a good service, and are considered high quality.

We have to say that the time for the compulsory education standards developing was sufficient, but that for textbook writing the period was too short. As mentioned above the trial draft standards document was issued in 2001 and the standards-based textbooks were put into use in some test areas in the same year. During the first 2 years of the test runs, the whole textbook series was still not ready. As a result, teachers did not have the next year's or the next two years' textbooks to consult. They were only informed of the outline of the other books. This made some teachers feel confused in planning their classes, especially for those topics designed in a spiral way. "Curriculum development should not be in haste" is one of strong criticisms of the MOE and it has been accepted.

Fan, Chen, Zhu, Qiu, and Hu conducted a study in China which investigated how teachers and students used textbooks within and beyond mathematics classrooms (Fan et al., 2004). They found that textbooks were the main source but not the only source for teachers to make decisions about what to teach and how to teach. For students, textbooks were their main learning resource for both in-class exercise and homework. They have provided a general picture of the textbooks used by Chinese teachers and students of mathematics. Here, I want to add some new findings reported recently which focus on the use of standards-based textbooks (Grades 7–9) (Yang, 2005).

In his master's thesis research, 76 schoolteachers and 862 students from 16 junior high schools in the region of Hefei in Jiangsu Province took the questionnaires. Two series of new standards-based textbooks were investigated in his study.

Before the reform, the PEP series were considered as classic textbooks, since they were modified time after time in the past 50 years. Although each series of new standards-based textbooks were modified year to year so that their quality improved, the majority of schoolteachers in his study believe examples and exercise problems need to be redesigned very carefully. Also, the rationale for big changes in content arrangement should be clarified. In an item asked "to what extent do you feel satisfied with the new standards-based textbooks," 46.0% of the teachers chose "basically satisfied," 14.5% of the teachers chose "satisfied" but none of them chose "very satisfied." In a similar item asked the students to what extent do they enjoy reading their textbooks, 31.2% of the students chose "enjoy basically," 37.8% of the students chose "enjoy," and 17.5% of the students chose "enjoy very much." It seems that the students had better attitudes toward the new textbooks than their teachers. Probably, it is a result of trying to make textbooks serve more as a learning resource than as a teaching resource, which is currently advocated by the MOE.

As for the role of these textbooks for teachers, Yang pointed out that the majority of the teachers depended highly on the textbooks, and they depended on textbooks in deciding what to teach more than in deciding how to teach. Teachers usually did not depend on textbooks in review lessons. Over 90% of the teachers admitted that they did add some contents not required by the new curriculum but appeared in the old textbooks, such as some important theorems, formulae, and examples. Teachers used almost all the textbook examples in their classroom teaching but they also selected examples from other teaching materials and exercise books. Sometimes they designed or modified examples by themselves.

It was found that "explore" and "reading material" were the two book features that the students liked most. "Project learning" was also a feature enjoyed by some of the students. However, these sections were not emphasized by the teachers. The conundrum of "teaching for exams" is still an unresolved issue in China. In order to help their students get high marks on examinations, most schoolteachers in China spend a lot of teaching time in lecture and doing exercises. They are usually reluctant to organize classroom activities that they think are not efficient in increasing students' scores. The voice of reform assessment of the system is strong, but time is needed for the changes to become effective.

In China, teachers usually work in a big office. They are grouped either according to the subject that they teach or the grade that they teach. As a result, teachers have many opportunities to discuss teaching and learning problems. Teachers teaching the same grade meet once a week. During these meetings, they usually share their teaching experiences and discuss their class plans for the next week. Teachers working in the same district also meet regularly and these meetings are chaired by teaching supervisors in the district. Classroom teaching observations, lectures, and teaching competitions are the main activities organized in the district. In addition, the Internet is another communication platform for some teachers. Consequently, teachers do not feel isolated or helpless when they face the new curriculum.

REFERENCES

Bao, J. (2004). A comparative study on composite difficulty between new and old Chinese mathematics textbooks. In L. Fan, N. -Y. Wong, J. Cai, & S. Li (Eds.), *How Chinese learn mathematics* (pp. 208–227). Singapore: World Scientific.

Cockcroft, W. H. (1982). *Mathematics counts (Report of the committee of inquiry into the teaching of mathematics in schools)*. London: H. M. S. O.

Fan, L., Chen, J., Zhu, Y., Qiu, X., & Hu, Q. (2004). Textbook use within and beyond Chinese mathematics classrooms: A study of 12 secondary schools in Kunming and Fuzhou of China. In L. Fan, N. -Y. Wong, J. Cai, & S. Li (Eds.), *How Chinese learn mathematics* (pp. 228–261). Singapore: World Scientific.

Fine, H. B. (1904). *College algebra*. Boston: Ginn.

Freudenthal, H. (1973). *Mathematics as an educational task*. Dordrecht, Holland: D. Reidel.

Goodlad, J. I., Klein, M. F., & Ty, K. A. (Eds.). (1979). The domains of curriculum and their study. In *Curriculum Inquiry* (pp. 43–76). New York: McGraw-Hill.

Granville, W. A. (1909). *Plane and spherical trigonometry, and four-place tables of logarithms*. Boston: Ginn and Company

Li, J. (2000). *Chinese students' understanding of probability*. Unpublished doctoral dissertation, National Institute of Education, Nanyang Technological University, Singapore.

Li, J. (2004a). Thorough understanding of the textbook: A significant feature of Chinese teacher manuals. In L. Fan, N. -Y. Wong, J. Cai, & S. Li (Eds.), *How Chinese learn mathematics* (pp. 262-281). Singapore: World Scientific.

Li, J. (2004b). Teaching approach: Theoretical or experimental?. In L. Fan, N. -Y. Wong, J. Cai, & S. Li (Eds.), *How Chinese learn mathematics* (pp. 443-461). Singapore, World Scientific.

Ma, Z. (2001). *History of mathematics education* (in Chinese). Nanning, China: Guangxi Education Press.

Ministry of Education of China. (2001). *The standards of mathematics curriculum for compulsory education (trial draft)* (in Chinese). Beijing, China: Beijing Normal University.

Ministry of Education of China. (2003). *The standards of mathematics curriculum for full-time senior high schools (trial draft)* (in Chinese). Beijing, China: People's Education Press.

Mo, Y., & Xu, Z. (1987). *History of modern mathematics in China* (in Chinese). Nanning, China: Guangxi Education Press.

National Research Council. (1989). *Everybody counts: A report to the nation on the future of mathematics education*. Washington, DC: National Academy Press.

Ngai, K. Y., Lee, C. S., & Ko, H. Y. (1987). *History of secondary mathematics education in China* (in Chinese). Beijing, China: People's Education Press.

Robitaille, D. F., Schmidt, W. H., Raizen, S., McKnight, C., Britton, E., & Nicol, C. (1993). *Curriculum frameworks for mathematics and science (TIMSS Monograph No. 1)*. Vancouver, British Columbia, Canada: Pacific Educational Press.

Schultze, A., Sevenoak, F. L., & Stone, L. C. (1935). *Plane geometry*. New York: Macmillan.

Schultze, A., Sevenoak, F. L., & Schuyler, E. (1928). *Solid geometry*. New York: Macmillan.

Smith, P. F., Gale, A. S., & Neelley, J. H. (1928). *New analytical geometry*. Boston: Ginn.

Sobel, M. A., & Maletsky, E. M. (1991). *Teaching mathematics*. Boston: Allyn & Bacon.

University of Chicago School Mathematics Project. (1992). *Functions, statistics, and trigonometry*. Glenview, IL: Scott, Foresman.

Valverde, G. A., Bianchi, L. J., Wolfe, R. G., Schmidt, W. H., & Houang, R. T. (2002). *According to the book. Using TIMSS to investigate the translation of policy into practice through the world of textbooks*. Dordrecht, The Netherlands: Kluwer Academic.

Wang, J. (2001). *Trial mathematics textbooks (for grade 7 to 9)*. Shanghai: East China Normal University.

Yang, W. (2005). *An Investigation of New Standards-based Mathematics Textbooks Use in Junior High Schools*. Unpublished master's thesis, East China Normal University, Shanghai.

CHAPTER 9

INNOVATIONS BRINGING DEGENERATION

A Lesson From Historical Analysis of the Revisions of the National Curriculum Standards for Upper Secondary School Math in Japan After World War II

Ryosuke Nagaoka
The University of the Air, Japan

This is a rough sketch drawn by a chief of the editorial board of a high school mathematics textbook series, of the roles the very strict National Curriculum Standards (MEXT 1998a, 1998b, 1998c, 1999) has played in Japanese school education, and a brief history of the revisions of the National Curriculum Standards for Upper Secondary School Mathematics after World War II, through which the author deduces the historical lesson that an "innovative" actions made in a "good will" do not necessarily attain the original aim, but sometimes end with unexpected results.

Mathematics Curriculum in Pacific Rim Coutries—China, Japan, Korea, and Singapore:
Proceedings of a Conference, pp. 141–154
Copyright © 2008 by Information Age Publishing

INTRODUCTION

The most frequently asked question by schoolteachers and mathematicians in Japan is:

Why does the National Curriculum Standards get worse and worse after repeated revisions?

It is very natural for schoolteachers to raise such a simple yet important and essential question. For them, revisions of the National Curriculum Standards appear to be periodic affairs which bring nothing good to anyone, but deliver more and difficult work to them. To most mathematicians, who refer to the National Curriculum Standards only when they are engaged in posing problems for the college entrance examination, the revisions appear to be carried out by an unknown force without reason. Surprisingly enough, no version of the National Curriculum Standards has been evaluated to be better or worse in any sense than former ones, although all of them were created and enacted through tremendous efforts by professionals, scholars, pedagogues, and practitioners after extensive discussions.

Although paradoxical, an unexpected effect can transform the result of an intention to the opposite of what was first embraced. I propose that an excessive naïve optimism concerning future innovations is the key factor diverting goodwill efforts to disaster and confusion in schools. With this I hope to attract readers' attention to a key point in implementing any kind of systemic standard for school curriculum, which I am afraid is often neglected or underestimated. To attain this aim the dialectical processes will be studied in a retrospective analysis of the efforts to "innovate" mathematics study in the Upper Secondary School.

JAPANESE SCHOOL SYSTEM AND THE
NATIONAL CURRICULUM STANDARDS

Before engaging the subject, a rough overview is given here. First I describe the school system and the character of the National Curriculum Standards in Japan, and second I describe the process whereby the National Curriculum Standards document is revised and enacted.

The Japanese school system before the university/college (tertiary) level consists of three stages:

Primary

1. Elementary school, or *Shougakkou*,[1] 6 years from first to sixth grades: for children from 6 to 12 years old.

Secondary

2. Junior high school, or *Chuugakkou*,[2] 3 years from seventh to ninth grades: for students from 12 to 15 years old

3. Senior high school, or *Koutougakkou*,[3] 3 years from 10th to 12th grades: for students from 15 to 18 years old.

Elementary school and junior high school are compulsory, while senior high school is not. However, the ratio of students who enter senior high schools today is very high, almost 100%, especially in metropolitan areas.

The most important aspect of the Japanese school system is the strong governmental control over curricula up to the secondary level. In fact, school curricula in Japanese schools are strictly required by law to be fully consistent with the National Curriculum Standards issued by the MOE (Ministry of Education).[4] This is achieved by demanding that, from the primary to the secondary level, every school must use MOE approved textbooks. This applies to all subjects and all schools, whether national, municipal, or private. Such textbooks are reviewed and approved for consistency with the National Curriculum Standards. (Of course, the decision procedures for textbooks are dependent on the school. Roughly speaking, private schools and schools for higher stages have more freedom in their choice and use.)

The MOE does not regard the approval process as a kind of inspection. Instead it is more of an equalizing process, in that no textbook is allowed to be of a higher or lower level or contain richer or poorer content with respect to the National Curriculum Standards. In this sense, there is only one school curriculum throughout all Japanese schools before the tertiary level. This is, of course, slightly exaggerated. Schools and schoolteachers, especially in private schools, can direct their own curricula which can be of a higher or lower level or contain richer or poorer contents at their own risk and responsibility. Still, they are not allowed to dispense with buying a textbook approved by the MOE. More importantly, the problems posed on the entrance examination of any university are strongly requested to be highly consistent with the National Curriculum Standards.

It might be difficult for a stranger to Japan to imagine the reason why the National Curriculum Standards controls the teaching in Japanese schools so strictly. In my view, the answer is quite simple: there has been no strong political leadership over school education since World War II.

Politicians have been interested in and anxious about matters considered more urgent and connected with big political and economic issues. Also, the two leading universities in the field of higher education, *KohtohShihan*[5] were abolished by GHQ[6] just after World War II. The graduates of these universities or the leaders of schoolteachers were regarded, with good reason, by GHQ as promoters and collaborators of militarism during the war.

This dearth of strong political or scientific leadership in education resulted in the MOE becoming the sole leadership with overwhelming power and responsibility. The MOE has regarded its chief mission to guarantee all kinds of equity in education throughout Japan. This includes equity in qualities of education, equity in contents of education, and sometimes even equity in the achievements of students! The MOE has a lot of channels, powers, and tools to exert influence upon schools and schoolteachers. But among all these, the National Curriculum Standards and the approval system for textbooks have been the key devices for the MOE to attain its mission.

THE WAY THE NATIONAL CURRICULUM STANDARDS IS REVISED

Here is described the way in which the National Curriculum Standards is revised and enacted. The National Curriculum Standards have been revised after World War II roughly every 10 years. One may ask: Why does it take so long time to revise the national curriculum? or Why should the national curriculum be revised so often? The answer to the first question is quite simple: it is the result of a lengthy democratic process in which no particular persons are responsible for the final decisions. With a lack of strong political guidance, any decisions are made through repeated lengthy discussions with committee members who are often too bored by the process to continue arguing for their own visions of change.

Every revision begins with top-level decision makers meeting to discuss educational policy. This is followed by somewhat more concrete discussions by scholars and practitioners. These in turn are followed by working group meetings where the final form is decided.

Any possible discrepancies between the new goals of school education, declared at the conclusion of the top-level meetings and the concrete embodiment of the National Curriculum Standards are not necessarily checked by any independent organization. A feasibility study of the new goals being attainable through the finalized curriculum standards is not made. Instead these possible problems are intentionally neglected and erased in an enthusiastic chorus of the new ideals held up by the MOE and its adherents.

Since the revision takes a long time and it covers all grades, 1–12, it usually takes a few years to transfer smoothly from the older version to the newer one. For example, some topics in the textbook are neglected and some topics are taught without using textbooks. However, in principle, no contradiction is allowed during this long process of enacting the standards.

In light of the constancy of the revision process, it might appear very surprising that there are no official reviews of past curricula. But there are no independent organizations to evaluate or assess the past National Curriculum Standards. Reviewing might be done, but it is done silently and only in the minds of those people who were engaged with a revision.

With few exceptions, there have been no special programs or streams in Japanese schools. This means that all students are to learn the same contents in the same grade in the same hours in accordance with the National Curriculum Standards.

THE NATIONAL CURRICULUM STANDARDS FOR TEACHERS

As the result of the "democratic" process of revision, with a little exaggeration, the work to revise the national standards begins just after it is enacted!

For most teachers the revisions of the National Curriculum Standards are as described above, just periodic affairs and not necessarily with high educational value. Nevertheless, the revisions of the standards have attracted the attention of schoolteachers. This is because the National Curriculum Standards substantially determines what should be taught and learned in schools, as well as what should be written in textbooks. The standards also have strong influence on the entrance exams for universities. Since preparation for these exams is very important, schoolteachers pay attention to the changes, especially through the secondary level.

Thus, the National Curriculum Standards has been the most important matter discussed among schoolteachers and educators, especially in the case of mathematics. As such, the National Curriculum Standards have attracted the keenest interest among schoolteachers due to the specific character of school mathematics: the most effective and decisive factor in the results of entrance examinations and the possible diversity of the quality of teaching and learning of material by students.

In this sense, the repeated revisions of the National Curriculum Standards for mathematics can be regarded as a significant driving force concerning schoolteachers and what they are to teach their students. In fact, without any revisions to the National Curriculum Standards, it is possible to say that some schoolteachers would have lost all their interest in mathematics as they think that they have mastered all of the content they are to teach. So called *lesson study* is quite common among schoolteachers, especially when a great deal of innovation is introduce with the new curriculum.

A SHORT HISTORY OF MATHEMATICS CURRICULA FOR THE SENIOR HIGH SCHOOLS AFTER WORLD WAR II

After World War II, the Japanese school system was largely changed. One of the most important changes enacted was separation of high schools into compulsory junior high schools and elective senior high schools.

While few changes were brought to the contents of mathematics taught at the compulsory levels of education, the National Curriculum Standards for senior high school mathematics was sometimes drastically revised. I will give a historical sketch of the revisions of the National Curriculum Standards. This will give a simple overview of mathematics education at the upper secondary level in Japan as well as a tentative understanding of the roles the National Curriculum Standards have played in the Japanese education system. This will also hint at key factors of real innovation in mathematics teaching and learning, which Japanese policymakers and their collaborators have underestimated for over half a century.

The first major revision of the National Curriculum Standards of mathematics for senior high schools was enacted for the class entering senior high schools in 1963, from hereon called the 1960s revision. The succeeding revisions came almost every 10 years and are referred hereto as the 1970s revision, the 1980s revision, the 1990s revision, and the 2000s revision respectively, each of which is discussed in detail below.

The 1960s Revision—The First Revision "Too Successful"

The National Curriculum Standards 1960s revision was the richest in mathematics content and the simplest in structure among all the later revisions. Senior high school mathematics was composed of only three subjects (or textbook units): math I for 10th grade, math II for 11th grade and math III for 12th grade. It was intended chiefly for students in the scientific fields.

1. Math I in the 1960s revision consisted of the following topics:
   - Numbers (irrational numbers, imaginary numbers) and formulas (polynomials and their fractions)

- Logic and sets (propositional and predicative logic, necessary and sufficient conditions, basic concepts of set theory)
- Quadratic equations/inequalities/functions
- Fractional/irrational equations/inequalities/functions
- Analytic geometry (lines, circles, quadratic curves, plane loci)
- Exponential and logarithmic functions/ equations/inequalities
- Trigonometry (metric problems)

2. Math II in the 1960s revision consisted of the following topics:
- Sequences and recursive functions
- Complex plane (Gaussian plane)
- Plane vectors and space vectors
- Combinatorics and probability
- Trigonometric functions (addition theorem and its applications)
- Differential calculus of polynomials and fractional functions
- Integral calculus of polynomials

Note that these were the contents included in the subject called math IIB, while there was another math II called math IIA. However, it was seldom studied by high school students who hoped to continue schooling after graduating from high school.

3. Math III in the 1960s revision consisted of the following topics:
- Limits of sequences (progressions) and infinite series, limits of functions (including the intermediate value theorem), continuity/discontinuity
- Differential calculus with elementary functions (including the mean value theorem)
- Integral calculus with advanced techniques (integration by parts, integration by substitution)
- Ordinary differential equations (chiefly of first order)
- Statistics (which was often neglected)

We can summarize the character of the 1960s revision as the systematic approach to calculus as the target of high school mathematics. But we should also pay attention to a small and modest step towards modernization in the revision, which can be seen in the introduction of the concept of a vector and the complex plane on the one hand and removal of classical synthetic geometry, which was one of the main themes of math I in the former National Curriculum Standards, on the other.

Another striking character of the revision was the richness of the content to be taught, especially in the first year of senior high school (10th grade in American sense). It was true that math I was too rich for most students even in those days.

But most important, the new National Curriculum Standards was accepted and supported by schoolteachers, students, parents, and people in general. That was because in the 1960s science and mathematics were well respected. Due to the social views, nobody could raise any objection against the promotion of mathematics education.

The 1970s Revision

The 1970s revision was an infamous "modernization" revision. Although the basic structures of math I, math IIB, and math III were kept almost the same as in the 1960s revision, many modernized topics were introduced from early stages: such as, conception of a mapping as a pure correspondence between two given sets in Dirichlet's style in math I. Abstract algebraic concepts, (algebraic operations and their properties) were taught in math I. Plane vectors were moved to math I, which had been a topic in math II in the 1960s. Axiomatic reconstruction of elementary geometry was taught in math IIB. Matrices and linear transformations were taught in math IIB, while study of the complex plane was removed from math IIB.

Clearly, in the 1970s revision, there was an idealism present in which high school mathematics would more closely resemble university mathematics. Modern mathematical concepts were unquestionably regarded as better materials than classical ones to teach and learn.

What an optimism it was!

Innumerable tragic and comic affairs resulted. Confusions, misunderstandings, and errors were noted although they were not reported officially. But this revision was not simply a failure made by modernizers and mathematicians with goodwill in promoting this new movement. It also caused a serious side effect throughout education. The 1970s revision brought a decrease in the credibility of schools and schoolteachers among parents. This was mainly because schoolteachers suddenly became less confident in their teaching, due to the shift in content of the new materials.

These problems were accelerated by the rapid economic growth of Japan and the closely related growth of the private tutoring industry, called *jukus*. Unfortunately, people's enthusiasm for and confidence in the sciences declined as a result of negative phenomena caused by technology, including environmental disasters. Soon, young competent students hoped to be medical doctors rather than scientists. In this sense, the revision was too optimistic also from a sociological point of view.

The 1980s Revision

The third revision, the 1980s revision, was written mindful of the severe criticism of the previous modernization movement. The most striking change of the National Curriculum Standards could be seen in the design of subject composition. The traditional system composed of three units, from math I to math III, was reconstructed into five subjects: math I, algebra and geometry, foundational calculus, probability and statistics, and differential and integral calculus.

Mathematical topics were reorganized, not according to their difficulty level, but according to their theoretical proximity. The traditional style of teaching apparently different materials belonging theoretically to the same topics repeatedly in different contexts, called the Spiral System, was abandoned. Although a few modern topics like the concept of mapping and the axiomatic reconstruction of geometry were removed, most components in the previous curricula were kept.

The 1980s revision appeared to be a moderate reaction to the modernization attempts of the previous revision, but actually it was not.

This reorganization of mathematics topics according to their theoretical proximity brought unexpected side effects. It made the textbook curriculum poorer. For example, integrated study of sequences and matrices (powers of matrices or iteration of linear transformations) or of sequences and probabilities (Markov chains), which had been very important topics in previous curricula, were forbidden topics in the textbooks approved by the MOE because sequences, matrices, and probability were topics in entirely different courses. Thus the refinement of mathematics topics which aimed at more efficient teaching, brought about the fragmentation or anti-integration of mathematics knowledge even among competent students.

Another serious side effect of the 1980s revision was the acceleration of a tendency towards diversity in university entrance examinations. The traditional single mathematics goal was abandoned because high school mathematics was decomposed into five independent subjects. Although only some of the five subjects came to be studied by a nonnegligible number of students, the style of students' study turned from the traditional diligent study for entrance examinations to the collection of information useful for particular individuals and correct decision of selection based on it.

The 1990s Revision

With an increasing ratio of students enrolled in senior high schools, a new scheme for designing school mathematics curricula was sought. In the 1990s revision, a new design philosophy for the National Curriculum Standards was born with the proposal of "core and option" and "supply on the spot." This was intended to bring increased flexibility to each school's curriculum by allowing a diversity of content to be taught and therefore diversity in goals and levels of learning.

In the 1990's revision, mathematics knowledge necessary for people to live in an age of advanced technology and information is given in core subjects, while advanced topics or remedial topics can be flexibly chosen based on the needs of students. But when the philosophy was implemented in the National Curriculum Standards, it brought the breakdown of the core knowledge of mathematics and actually no flexibility.

With the introduction of core and option the contents of high school mathematics were divided into math I, math II, and math III, forming the core, and math A, math B, and math C, forming the options.

Core:
- Math I was composed of four topics: (1) Quadratic functions and inequalities, (2) progressions of integers, (3) trigonometry, and (4) probability.

- Math II was composed of four topics: (1) Analytic geometry restricted to lines and circles, thus excluding parabola and other conic sections; (2) exponential and logarithmic functions; (3) trigonometric functions, restricted to angles from $0°$ to $360°$; and 4) differential calculus of polynomial functions of less than fourth degree and integral calculus of polynomials of less than second degree (volume calculus is excluded).

- Math III was composed of the following four topics: (1) Abstract concept of functions, (2) limits of sequences and functions, (3) differential calculus, and (4) integral calculus.

Options:
- Math A was composed of the following four topics: (1) Numbers and formulas (imaginary numbers and fractions of polynomials are excluded), (2) sequences (progressions) and recursive definitions, (3) elementary computer programming, and (4) elementary synthetic geometry.

- Math B was composed of four topics: (1) Vectors (in two and three dimensions), (2) complex numbers and quadratic equation (including complex plane), (3) random variables and their distributions, and (4) the use of application software in mathematical contexts.

- Math C was also composed of four topics: (1) Curves, (2) matrices, (3) statistics, and (4) advanced computing.

In the optional subjects, schools and teachers were allowed to choose any two topics among the four. However, this process did not work well since the mathematical and educational weights of the four topics were not at all equivalent. For example in math A, almost all teachers chose numbers and formulas and sequences and recursive definitions instead of elementary synthetic geometry or computer programming.

Moreover, to make the core subject as simple and as basic as possible, the most fundamental subject, math I was composed only of four basic topics as shown above. All complicated elements included in the older version of math I were removed, while trigonometry and probability were kept because they were mathematical topics to be found in real life. However, simple progressions of integers was expected to attract students' interest in mathematics.

As a result of positioning numbers and formulas as a topic in an optional course, math A, one could not assume that students had even the most basic knowledge associated with this topic. This in turn unexpectedly brought about excessive simplification of many other topics. This brought the most serious effect, the decay of essential knowledge and basic skills among students, which soon appeared very serious to all who were engaged in education.

The 2000s Revision

After the political decision of the transition to five school days per week, the contents of mathematics at the primary level (elementary school) and lower secondary level (junior high school, Grade 7-9) were reduced by about 30%.

As a result of these reductions, some important topics were moved from the junior high school curriculum to the senior high school curriculum. Nevertheless, most topics in the former curricula were preserved in order to "keep the goal of high school mathematics education!" The school curriculum was thus obliged to be too condensed in the 3 years of upper secondary school.

The popularity of private high schools, which have junior high and senior high in them, is on the rise, especially in metropolitan areas. Many people think it is better to be continuously in the same school than to attend different junior and senior high schools. Even among public schools, there has begun a new challenge of reorganizing the high school system with combined junior and senior high. The 2000s revision will surely accelerate the trend, because the combined school approach may allow for greater attention to the problems this overly condensed curriculum brings.

What spurs the establishment of this overly condensed curriculum in the 2000s revision? One cause is the general opinion of teachers and mathematicians that younger people have failed to develop fundamental skills and knowledge during the unhappy ten years of the previous curriculum. Thus, any further reductions in content would be disastrous. Also, the MOE has recently decided to change a long-standing policy of maintaining an equal level of teaching and learning throughout Japan. The MOE now says that the National Curriculum Standards should be regarded only as a required minimum. Moreover, most private high schools have returned to the traditional 6 days per week structure.

Considering the traditional way of thinking of teachers and people raised with the equity maintaining policy of the MOE, it is not easy for everyone to accept new conditions readily. Certainly, much more confusion will develop, not just within mathematics education, but throughout the entire school system.

Anyhow, one can see here also a kind of paradox, in that the efforts to keep the goals of high school mathematics education stable, bring about the destruction or rather the reorganization of high schools.

WHAT IS NEEDED FOR A BRIGHTER FUTURE?

From my simple analysis of the history of revisions of the National Curriculum Standards for senior high school (Grade 10-12) mathematics in Japan, one can find the real causes which made mathematics education worse.

I dare to sum up the important points very concisely. First, the implementation of the goodwill of mathematicians leads to unintended learning experiences: often "new mathematics with more joy" is actually "strange mathematics, just with more pain." This of course can be avoided if the contents are very carefully selected and organized.

Second, the optimistic view about the implementation of curriculum often takes into account only the brightest students and best teachers. This might not be optimism but an opportunism.

Third, a stubbornness and conservativism among teachers prevents implementation of many of the intended revisions. Most teachers' knowledge of math is the same as they had in their high school days. In Japan, there are no systematic training programs for individual teachers in service. But anyway it is the strict boundary condition within which we should seek the best solution. We cannot neglect it!

What is needed to realize a better future for education in Japan is wisdom among mathematicians against irresponsible optimism and confident teachers not only with teaching experience but also with the love of mathematical thinking. This it is easy to point out but difficult to realize!

However, we can do something positive or at the very least we can avoid doing something negative to achieve the ideal goal. Those in charge of revisions should be a little more conservative and or modest in reforming curricula, so that we can expect schoolteachers to be more confident of their knowledge of mathematics and so that what new material needs to be learned is not too great. Confidence in a subject is the most essential key of quality teaching.

In my view, mathematical teaching in the primary level has been relatively successful in Japan. I assume several reasons for that, but among them, the decisive one being that relatively fewer and smaller changes in the mathematical contents of curricula have been made in the revisions of the National Curriculum Standards at the primary level. Hence, Japanese elementary school teachers have confidence in their knowledge of the mathematics they teach. This will be a good hint to consider what the ideal curriculum standards should be. Anyway, an innovative attitude is not always truly innovative! The most important point always to be taken into consideration is that students have a right to receive their education in mathematics from confident teachers who love mathematics!

Finally, needless to point out for non-Japanese people, allow me to add one last remark: What is most important is to decide the kind of education and the range of flexibility allowed in that education as well as designing an ideal curriculum. It is undoubtedly true that Japan has been very successful in building human capacity through high school education, at least previous to the past 2 decades. To some degree it was made possible through the National Curriculum Standards and quality textbooks strictly consistent to it. But to find the harmony between the normative standards and its flexible application is not trivial.

NOTES

1. *Shou* means "small" and *gakkou* means "school."
2. *Chuu* means "middle."
3. *Koutou* means "high" or "higher."
4. The MOE, being combined with the Ministry of Science and Technology, has recently been reorganized as the Ministry of Education, Science and Technology.

5. *Kohtoh* stands for "higher" just as in the case of *KohtohGakkou,* and *Shihan* means "teachers." In this sense, KohtohShihan can be compared to the French *Ecole Normale* established after the French Revolution.
6. General Headquarters of the Supreme Commander for the Allied Powers

REFERENCES

Ministry of Education, Culture, Sports, Science, and Technology. (1998a). *Yochien Kyoiku Yoryo* [Curriculum standards for kindergarten]. Tokyo: Printing Bureau.

Ministry of Education, Culture, Sports, Science, and Technology. (1998b). *Shougakkou Gayusyu Sido Yoryo* [Curriculum standards for elementary school]. Tokyo: Printing Bureau.

Ministry of Education, Culture, Sports, Science, and Technology. (1998c). *Chuugakkou Gakusyu Sido Yoryo* [Curriculum standards for junior high school]. Tokyo: Printing Bureau.

Ministry of Education, Culture, Sports, Science, and Technology. (1999a). *Koutougakkou Gakusyu Sido Yoryo* [Curriculum standards for high school]. Tokyo: Printing Bureau.

PART III

U.S. PERSPECTIVES ON THE
CURRICULA OF ASIAN PACIFIC RIM COUNTRIES

SOME HIGHLIGHTS OF THE SIMILARITIES AND DIFFERENCES IN INTENDED, PLANNED/IMPLEMENTED, AND ACHIEVED CURRICULA BETWEEN CHINA AND THE UNITED STATES

Jinfa Cai
University of Delaware

It is widely accepted that a major goal of educational research is to improve learning opportunities for all students. In order to improve students' learning, it is necessary to understand the developmental status of their thinking and reasoning. Thus, the more information teachers obtain about what students know and think, the more opportunities they create for student success (Darling-Hammond, 1994). Teachers' knowledge of students' thinking has a substantial impact on their classroom instruction, and hence, on students' learning (Fennema & Franke, 1992; Gardner,

Mathematics Curriculum in Pacific Rim Coutries—China, Japan, Korea, and Singapore: Proceedings of a Conference, pp. 157–181

1999; Wittrock, 1986). Cross-national comparisons of curriculum and instructional practices provide a unique perspective for understanding what students know and think, as well as seeing how we should help students learn mathematics with understanding. Such an international perspective can increase educators' and teachers' experiences when they try to address the issues and challenges facing students' learning of mathematics with understanding (Cai, 2001; Gardner, 1989; Ma, 1999; Stigler & Hiebert, 1999). In other words, cross-national comparisons of curriculum and instruction not only can reveal what topics are or are not treated in different curricula, but they also can show how the same topics are treated.

In the past decade, I have engaged in such a cross-national comparative project. Figure 10.1 shows the framework for the project that guided a series of investigations. In this project, we first examined the "achieved curriculum," which is what the students actually learn. Through assessing

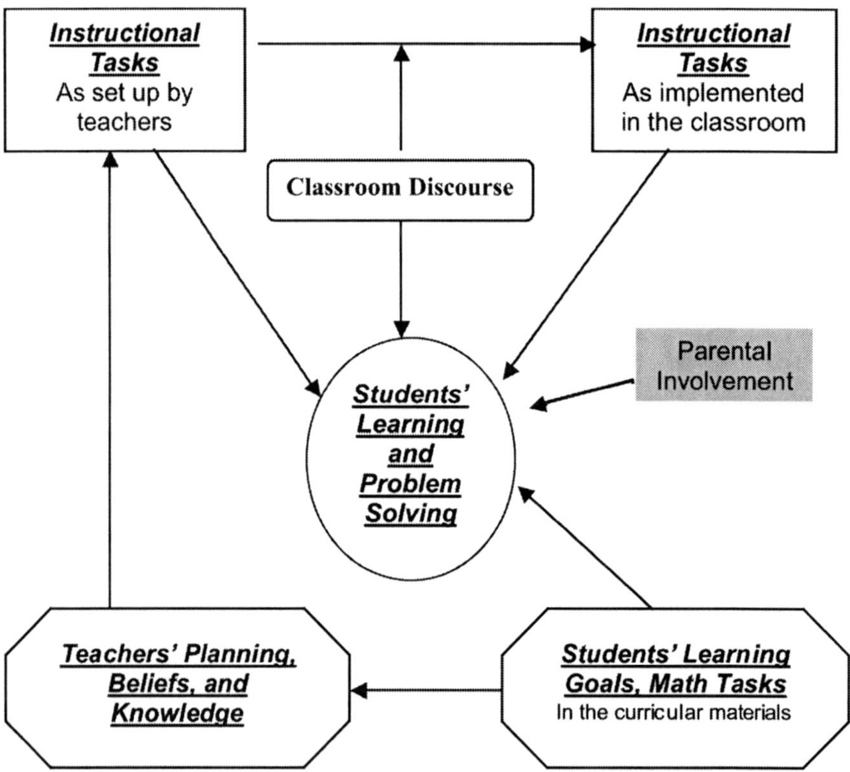

Figure 10.1.   A framework for cross-national comparative studies.

Chinese and U.S. students' mathematical problem solving and problem posing, these studies revealed many important and interesting similarities and differences between Chinese and U.S. students' mathematical thinking (e.g., Cai, 1995, 2000a, 2000b; Cai & Hwang, 2002). We also examined the "intended curriculum," which details what mathematics students need to know and what teachers are to do to help students develop their mathematical knowledge. In particular, we examined intended treatment of Chinese and U.S. curricula on various important mathematical topics, including arithmetic average, ratio and proportion, and early development of algebraic thinking (e.g., Cai, 2004a, 2004b; Cai & Sun, 2002; Cai, Lo, Watanabe, 2002; Cai et al., 2005). We examined both what Chinese and U.S. teachers actually teach in the classroom and how they plan what they teach and how they teach it in classroom (planned/implemented curriculum) (e.g., Cai, 2004, 2005, 2006; Cai & Wang, 2006). In addition, we examined the parental roles in Chinese and U.S. students' learning of mathematics in the home setting (e.g., Cai, 2003).

The purpose of this paper is to highlight some of the curricular similarities and differences between China and the United States. I will highlight these similarities and differences using three aspects: (1) U.S. and Chinese students' performance on various tasks; (2) U.S. and Chinese teachers' conceptions and construction of representations in classrooms; and (3) U.S. and Chinese intended curricular treatment.

U.S. AND CHINESE STUDENTS' MATHEMATICAL PERFORMANCE ON FOUR TYPES OF TASKS

In a study (Cai, 2000a), Chinese and U.S. sixth-grade students' mathematical performances were examined in four types of assessment tasks: (1) 13 multiple-choice tasks measuring computation skills, (2) 18 multiple-choice tasks measuring simple problem-solving skills, (3) six process-constrained, performance assessment tasks measuring complex problem-solving skills, and (4) six process-open, performance assessment tasks measuring complex problem-solving skills. A process-constrained task requires carrying out a procedure or a set of routine procedures in solving the problem. On the other hand, a task that is process-open may not require the execution of a procedure or a set of procedures, instead it requires exploration of the problem situation and then the problem can be solved. Appendix A shows a process-constrained performance assessment task (Hats Average Problem) and a process-open performance assessment task (Odd Number Pattern Problem).

Figure 10.2 shows the percent mean scores of the Chinese and U.S. students on the computation tasks, simple problem-solving tasks, process-

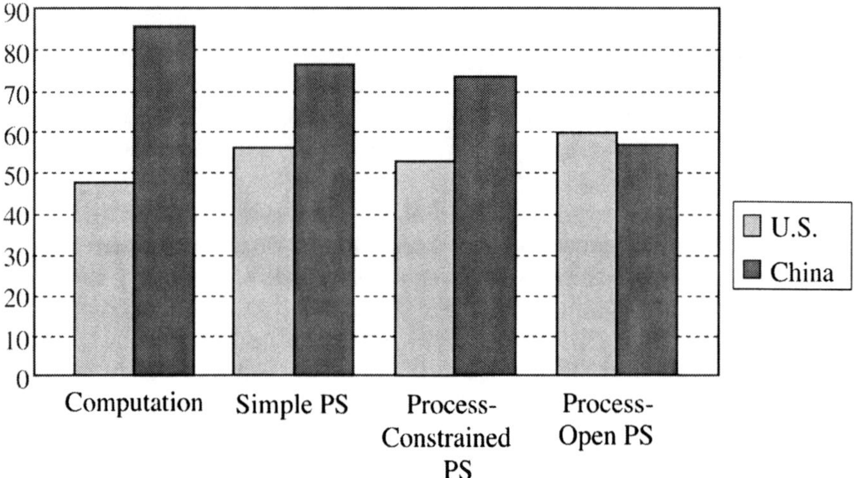

Figure 10.2.   Chinese and U.S. students' mathematical performance on four types of tasks.

constrained performance assessment tasks, and on the process-open performance assessment tasks. The Chinese sample had significantly higher mean scores than did the U.S. sample on the computation tasks, simple problem-solving tasks, and on the process-constrained tasks. However, the U.S. sample had significantly higher mean score than did the Chinese sample on the process-open performance assessment tasks. For the U.S. sample, the mean score on the process-open tasks was the highest, while the mean score on the computation tasks is the lowest. For the Chinese sample, however, the mean score on the process-open tasks is the lowest, while the mean score on the computation tasks was the highest.

While it is useful to know the performance differences in terms of mean scores or average correctness, we also need to describe student performance cognitively or qualitatively to understand the nature of the international differences in students' mathematical thinking and reasoning. In addition to examining the correctness of Chinese and U.S. students' solutions to each performance assessment task, we employed a qualitative analysis scheme to understand the nature of the differences (Cai, 1997). This qualitative analysis revealed that Chinese and U.S. students used different strategies and representations. For example, four categories were used to evaluate and classify representations for student's responses to the Hats Average Problem: verbal (primarily written words), pictorial (a picture or drawing), arithmetic (arithmetic expressions), and algebraic (algebraic expressions). Table 10.1 shows the percentages of

U.S. and Chinese students using each of the representations. The Chinese students used symbolic representations more frequently than did the U.S. students; however, U.S. students used verbal and pictorial representations more frequently than did Chinese students. Nearly one fifth of the Chinese students used algebraic representations and 76% of them used arithmetic representations. Nearly 60% of the U.S. students used arithmetic representations, but only one U.S. student used an algebraic representation. Nearly 10% of the U.S. students used pictorial representations, but none of the Chinese students used pictorial representations in their explanations. Appendix A shows an example of a pictorial representation (Hats Average Problem, Response B).

It was also found that the representations used are related to the success on the problem-solving measure using the other 10 open-ended tasks. As shown in Table 10.2, for the Chinese sample, students who used algebraic representations performed better than those who used any other representations. U.S. students who used arithmetic representations performed significantly better than those who used verbal and pictorial representations on the ten open-ended problems. Overall, Chinese students performed better than U.S. students on the 10 open-ended problems. However, if the analysis is limited to U.S. students using symbolic

Table 10.1.   Percentages of U.S. and Chinese Students in Each Representation for the Hats Average Problem

	Representations			
	Algebraic	*Arithmetic*	*Pictorial*	*Verbal*
United States (n = 229)	0[*]	59	8	33
China (n = 298)	19	76	0	4

Note: *One U.S. student used an algebraic representation. The zero percent is due to rounding.

Table 10.2.   Mean Scores on the Ten Open-Ended Tasks in Each Representation Related to the Hats Average Problem

	Algebraic *U.S. = 1;* *CH = 58*	*Arithmetic* *U.S. = 135;* *CH=227*	*Pictorial* *US = 18;* *CH = 0*	*Words* *U.S. = 75;* *CH = 13*
U.S. students (n = 229)	31.00	25.17	17.18	19.64
Chinese students (n = 298)	26.29	24.96	N/A	24.92

(algebraic or arithmetic) representations for the Hats Average Problem, there is no mean difference between Chinese and U.S. students' performance on the 10 open-ended tasks. This finding suggests that Chinese students' superior performance on the Hats Average Problem may be due, in part, to their use of more sophisticated representations (e.g., algebraic). Similar results were found in other studies (e.g., Cai & Hwang, 2002)

It should be noted that while Chinese students performed much better than the U.S. students in solving the Hats Average Problem, the two groups of students committed similar errors, which were mainly due to their "incorrect use of the computational algorithm." There were six incorrect ways students applied the averaging algorithm, which are described below:

1. 1.The student added the number of hats sold in week 1 (9), week 2 (3), and week 3 (6), then divided the sum by 3, and got 6. However, the average was 7. Therefore, the student added 3 to the sum of the numbers of hats sold in the first three weeks, then divided it by 3, and got 7, and then gave the answer 3.

2. The student added the number of hats sold in week 1 (9), week 2 (3), and week 3 (6), then divided the sum by 3, and got 6, 6 + 1 = 7. So the student gave the answer 1.

3. The student added the number of hats sold in week 1 (9), week 2 (3), and week 3 (6), then divided the sum by 3. The student then gave the quotient (6) as the answer.

4. The student added the number of hats sold in week 1 (9), week 2 (3), week 3 (6), and the average (7), then divided the sum by 4. The student then gave the whole number quotient (6) as the answer.

5. The student added the number of hats sold in week 1 (9), week 2 (3), and week 3 (6), then divided the sum by 4. The student then gave the quotient (4.5) as the answer.

6. The student added the number of hats sold in week 1 (9), week 2 (3), and week 3 (6), then divided the sum by 7. The student then gave the whole number quotient (2) as the answer.

While students' solution strategies can be examined in terms of abstractness versus concreteness as in the Hats Average Problem, they can also be examined in terms of mathematical conventionality. A conventional strategy is one that is usually taught in the classroom; in contrast, an unconventional strategy may not necessarily be taught in the classroom and may evolve from the students' novel explorations. For some tasks, students' solution strategies are better examined in terms of their abstractness; for others the degree of conventionality exhibited by a student's strategies better captures the underlying reasoning. The results from a number of studies showed that Chinese students are more likely to use conventional strategies. For example, when U.S. and Chinese sixth-grade students were asked to solve a Pizza Ratio Problem (see Appendix A), in which they needed to determine if each girl or each boy gets more pizza when seven girls share two pizzas and three boys share one pizza equally, they used eight different ways to justify that each boy gets more than each girl (Cai, 2000a). For those who used appropriate strategies, over 90% of the Chinese students used the following conventional strategy: *Each boy will get 1/3 of a pizza and each girl will get 2/7 of a pizza. If you compared 1/3 with 2/7, you would know that 1/3 is bigger than 2/7 by transforming them into common fractions (1/3 = 7/21 and 2/7 = 6/21. 7/21 - 6/21 = 1/21) or decimals (1/3 = .33 and 2/7 = .29. .33 - .29 = .04).* However, only about 20% of the U.S. students used such a conventional strategy. In contrast, the vast majority of the U.S. students used one of the following non-conventional strategies.

Solution 1:   Three girls share one pizza, and another three girls share another pizza. Each of these six girls will get the same amount of the pizza as each of the three boys. But one of the girls has no pizza. So, each boy will get more.

Solution 2:   Three girls share one pizza and the remaining four share one pizza. Each piece that each of the remaining four girls get is smaller than the boys get. So the boys get more.

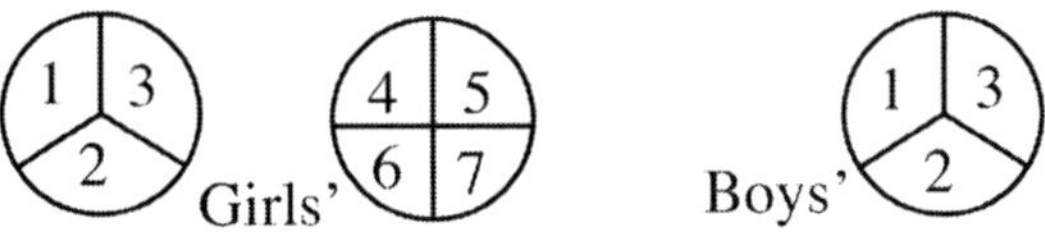

Solution 3:   7 Girls get two pizzas, and 3 boys get one pizza. The girls have twice as many pizza as boys. But the number of girls is more than twice as many than boys. So the boys get more.

Solution 4:   Each pizza was cut into 4 pieces. Each girl gets 1 piece with 1 piece left over. Each boy gets 1 piece with 1 piece left over.

The one piece left over must be shared by the 7 girls, but the 1 piece left over will be shared by three boys. So the boys get more.

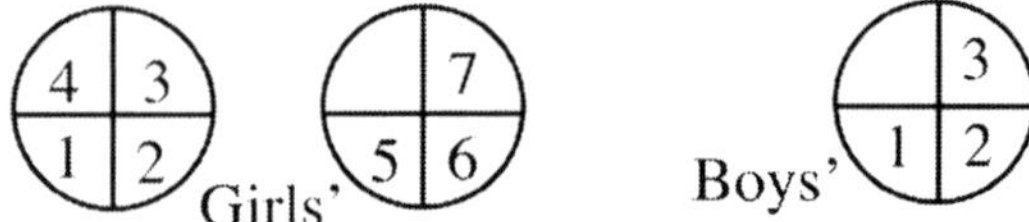

This example shows a dilemma we face. Apparently, the conventional strategy is quite efficient, and it can be easily applied to solve other similar problems, but this conventional strategy shows little originality. While these unconventional strategies show the originality of students' thinking, they also are task specific and less applicable for solving other similar problems, especially those that involve bigger numbers. The results from this particular example may suggest the effectiveness of Chinese classroom instruction on developing students' efficient strategies and the effectiveness of U.S. classroom instruction on developing original mathematical thinking. Ideally, we would hope that instruction can both foster students' learning of efficient problem solving strategies and develop their mathematical thinking with originality. If that is one of the goals for school mathematics, we have to seriously investigate the classroom instruction in both nations so that each can benefit from the other.

U.S. AND CHINESE TEACHERS' CONCEPTIONS AND CONSTRUCTIONS OF REPRESENTATIONS

Previous studies have revealed remarkable differences between Chinese and U.S. students' mathematical thinking and reasoning (Cai, 2000a; Cai & Hwang, 2002). Yet, we are just beginning to uncover factors that may contribute to our understanding of cross-national performance differences. Although there is no universal agreement as to whether mathematics is a culturally bound subject, no one questions the idea that the teaching and learning of mathematics is a cultural activity (Bruner, 1996; Stigler & Hiebert, 1999). In this section, we highlight some findings about Chinese and U.S. teachers' planned/implemented curricula. In particular, this section presents findings related to Chinese and U.S. teachers' conceptions and constructions of representations in mathematics instruction.

In one of the studies, a group of nine Chinese experienced teachers and a group of 11 U.S. experienced teachers were interviewed to examine their conceptions and constructions of pedagogical representations (Cai, 2005, 2006; Cai & Wang, 2006). They were asked to write lesson plans on the same topics, predicting students' possible solution strategies and representations, and evaluating students' representations and solution strategies.

PREDICTING STUDENTS' SOLUTION STRATEGIES AND REPRESENTATIONS

Table 10.3 shows the approaches U.S. and Chinese teachers thought their students would use to solve each of the five problems involving arithmetic average (see Appendix A). Overall, the Chinese and U.S. teachers' predicted solution strategies and representations are quite consistent with what Chinese and U.S. students actually used, found in Cai (2000a). In fact, all nine Chinese teachers thought that their students would use an algebraic approach to solve the Hats Average Problem. U.S. Teacher 9 was the only teacher to predict that students would use an equation-solving approach to solve the Hats Average Problem:

> *…They would be to add up what they already had. Well, they would have already added that up. Then they would be to say that we know what we're going to have to be divided by. There's four weeks, we're going to have to divide by four. So we're going to divide something by four, but we don't know, so I put a question mark. But we know that we want our answer to be seven.*

Chinese Teacher 9 also predicted that some of her students might use the following algebraic approach to solve the Score Problem: *Let x be the new average for the remaining eight scores. Then $8 \times x + 95 + 55 = 10 \times 87$. $x = 90$.* None of the U.S. and Chinese teachers thought their students would use the algebraic approach to solve the Elevator Average Problem, Cost Average Problem, and Baby Average Problem. These predictions seem to be natural because an algebraic approach is not very accessible for solving these problems.

An arithmetic approach is a popular strategy for both the U.S. and Chinese teachers. All the Chinese teachers and a considerable number of the U.S. teachers thought that their students would use arithmetic approaches to solve four of the problems. For example, these teachers predicted that their students would calculate and compare the average number of babies per litter for each mouse and each cat to solve the Baby Average Problem. For the Cost Average Problem, no teacher said that students would use an arithmetic approach.

Nine U.S. teachers, but only one Chinese teacher, predicted a drawing approach for the Hats Average Problem. Following is an example from U.S. Teacher 8, who predicted a drawing strategy for solving the Hats Average Problem:

Table 10.3.   U.S. and Chinese Teachers' Predicted Solution Strategies

Approach Tasks	Algebraic Approach	Arithmetic Approach	Drawing Approach	Guess & Check	Estimation	Incorrect Approaches
Hats	All Chinese teachers US9	All Chinese teachers, US1, US5, US6, US9, US10	CH7, US1, US2, US3, US4, US7, US8, US9, US10, US11	US1, US2, US3, US4, US5, US6, US8, US10		CH5, US2, US3, US10
Score	CH9	All Chinese teachers, US1, US2, US4, US5, US6, US7, US10, US11		US3, US4, US8, US11		CH3, US1, US2, US3, US6, US11
Elevator		All Chinese teachers, US1, US2, US3, US4, US5, US6, US7, US9	US3, US8, US10, US11			US1, US2, US3, US5, US6, US7, US8, US9, US10, US11
Cost			US3, US9, US11	CH1, CH2, CH4, CH5, CH6, CH8, CH9, US1, US5, US6	CH1, CH2, CH3, CH7, CH8, CH9, US1, US2, US4, US7, US8, US9, US10, US11	CH1, US2, US3, US8, US11
Baby		All Chinese teachers, US1, US5, US6, US7, US9	US1, US2, US3, US4, US8, US9, US10, US11		CH 3, CH5, CH6, CH7, CH9, US1, US2, US3, US4, US7, US8, US9, US10, US11	CH3, CH4, CH7, CH 8, US1, US3, US4, US6, US8, US9, US11

I think that some of my kids would figure out how many they had ... no, I think some of my kids would put seven in here. And I think they would take a couple from here and try to make seven. Uh huh. So they would like, one, two, three, four, five, six, seven, and they would chop these two off and put these two there. And say that doesn't make seven, so I must have to add

Table 10.4.   Rank Orders of U.S. and Chinese Predicted Difficulties for the Problems

Tasks	CH1	CH2	CH3	CH4	CH5	CH6	CH7	CH8	CH9	Mean
Hats	5	4	3	4	5	5	4	5	5	4.44
Elevator	4	5	5	5	4	4	5	4	4	4.44
Score	1	2	2	1	1	3	1	1	2	1.56
Book	2	1	1	2	3	2	2	3	1	1.89
Baby	3	3	4	3	2	1	3	2	3	2.67

Tasks	US1	US2	US3	US4	US5	US6	US7	US8	US9	US10	US11	Mean
Hats	5	5	3	4	5	3	4	3	3	5	3	3.91
Elevator	2	2	2	1	1	1	2	1	2	2	1	1.55
Score	1	1	1	2	2	2	1	2	1	1	2	1.45
Book	4	4	5	5	4	4	5	5	5	3	5	4.45
Baby	3	3	4	3	3	5	3	4	4	4	4	3.64

Note: "1" means the most difficult and "5" means the least difficult.

Table 10.5.   U.S. and Chinese Teachers' Scoring of the Six Responses

Responses	CH Teachers (n = 59)	US Teachers (n = 52)
A	3.73 (0.715)*	3.77 (0.469)
B	3.17 (1.020)	3.85 (0.364)
C	3.73 (0.739)	3.63 (0.595)
D	3.88 (0.326)	3.77 (0.469)
E	2.78 (1.001)	2.96 (1.028)
F	1.92 (1.368)	2.43 (0.962)

Note: *Numbers in parentheses are the standard deviations.

more. And they'd put a couple more on and then take them and put them there. And then when they had seven, they'd figure out what they had to put in there.

Eight of the U.S. teachers, but none of the Chinese teachers, thought their students would use a drawing strategy to solve the Baby Problem. For both the Hats Problem and Baby Problem, pictures, which may invite drawing strategies, were shown. Some U.S. teachers even predicted that their students might use a drawing strategy to solve the Cost Problem even though a drawing strategy is not necessarily apparent. For example, for the Cost Problem, U.S. Teacher 3 drew 10 books and put a price on each of the books. Finally, she subtracted some money from one book and added that amount to another in order to make the average $24.

RANKING DIFFICULTY LEVELS

After the teachers completed their prediction of students' solution strategies, each of them was asked to rank the five problems' in order of difficulty (see Table 10.4). There is an extremely high internal consistency for both groups of teachers. In fact, Cronbach's alpha is .973 for the U.S. teachers and .961 for the Chinese teachers. However, the correlation coefficient between the U.S. and Chinese teachers' mean ranking is only −.048. This result suggests that the overall ranking across the five problems is remarkably different between the U.S. and Chinese teachers. While solving the Hats Problem, Elevator Problem, and Score Problem requires a flexible application of the averaging algorithm, solving the Book Problem and Baby Problem requires an appropriate interpretation and use of the mean in a statistical context. It is clear that the two problems requiring an appropriate interpretation and using the mean in a statistical context are more difficult for Chinese teachers than for U.S. teachers. For the three problems requiring a flexible application of the averaging algorithm, the results are mixed. Both the U.S. and Chinese teachers thought the Score Problem would be the most difficult problem for their students. For the U.S. teachers, the Book Problem would be the easiest for their students, but for Chinese teachers, the Hats Problem and Elevator Problem would be the easiest. On the other hand, the Book Problem was the second most difficult for the Chinese teachers and the Elevator Problem was the second most difficult for the U.S. teachers.

EVALUATING STUDENTS'
SOLUTION REPRESENTATIONS AND STRATEGIES

In another study (Cai, 2004a), a group of 59 Chinese teachers and 52 U.S. teachers were asked to evaluate a set of 28 students' responses. Of the 59 Chinese and 52 U.S. teachers, nine of the Chinese and 11 of the U.S. teachers were interviewed. Table 5 shows the mean score for the six student responses to the two problems involving the arithmetic mean (Hats Average Problem and Score Average Problem). These six student responses are described in Appendix A. In Responses A, C, D, and E, "conventional strategies" were used, and in Responses B and F, "unconventional strategies" were used. For responses using conventional strategies, both the U.S. and Chinese teachers gave them similarly high remarks. The differences existed in the U.S. and Chinese teachers' scoring of responses with non-conventional strategies. In particular, for Response B, the U.S. teachers scored it higher than did Chinese teachers. For the majority of U.S. teachers, using drawing to even out the number of hats sold in four weeks was a viable solution strategy; thus Response B was scored as 4. For most of the U.S. teachers, this visual approach clearly showed how students thought about average and how they solved the problem. However, the Chinese teachers not only scored it lower than did the U.S. teachers, but the range of their scores was also greater than that of the U.S. teachers. Some Chinese teachers felt that the drawing strategy to even out the number of hats sold in four weeks was not as good as the strategies used in Responses A or C because drawing strategies were less generalizable. The Chinese teachers gave lower scores to Response B because "It is difficult to solve for larger numbers."

It is worth noting that Chinese teachers seem to hold two different views regarding Response B. For some of the Chinese teachers, the drawing strategy through leveling is difficult when solving similar problems involving larger numbers, so it should not have been given 3 or 4 points. However, for others, this approach showed the understanding of the averaging process, so it should be given 4 points. The Chinese teachers seemed to have a clear goal: students should learn more generalized strategies. The following excerpt from Teacher 7 is just one example showing that Chinese teachers have such a goal: *Being able to solve a problem is good, but just the first step. Through mathematics instruction, we want students to learn generalized problem-solving methods. They should be able to make a generalization and transfer it to other problem situations.*" However, there is no evidence from the interviews that U.S. teachers had such a clear goal. Instead, the U.S. teachers' goal was to have students solve a problem no matter what strategies they used. In fact, two of the U.S. teachers explicitly stated that as long as their students were able to use an appropriate strategy to solve a problem, they would be satisfied.

For both the Chinese and U.S. teachers, Response F was the most challenging one to evaluate because it is an unconventional response. There were quite large variations in both the U.S. and Chinese teachers' scores of Response F. The range of the scores for the U.S. teachers was from 0 to 3. The Chinese teachers scored the response either 0 or 2. Apparently, the teachers hold differing views about accepting an estimate as an answer. Some U.S. teachers gave Response F 2 or 3 points because they believed that students understood the problem and used some properties of arithmetic mean to find an answer. For example, U.S. Teacher 10 commented: *"They could do better because of the information given to them, but I guess with-*

out using all the information they've attacked the problem well. This student proved an understanding of the concept." Other teachers gave it 0 point or 1 point because they did not feel an estimate constituted an acceptable answer when students had enough information to find an accurate answer. For example, Chinese Teacher 4 said: "*I don't like this (Response F). They were provided all of the givens, and there is no need to estimate. This is a wrong answer for this problem.*"

ANALYSIS OF VIDEOTAPED LESSONS AND LESSON PLANS

The analyses of the videotaped Chinese and U.S. lessons showed a strong correlation between students' use of solution representations and teachers' use of pedagogical representations (Cai & Lester, 2005). The teachers in U.S. classrooms were much more likely to use concrete visual representations than were the teachers in Chinese classrooms. Analysis of Chinese and U.S. teachers' lesson plans showed similar findings (Cai, 2005; Cai & Wang, 2006). Even for the lesson where most of the U.S. and Chinese teachers used manipulatives, the function of the manipulatives in the Chinese and U.S. lessons are different. For the Chinese teachers, the main purpose of using visual models is to mediate students' problem solving and to support learning the averaging algorithm, but the visual representations themselves are not part of the solutions to mathematical problems. Therefore, Chinese teachers made explicit connections between the visual representations and the averaging algorithm.

Similarly, manipulatives and physical measurement activities are used to model the process of finding the mean in some of the U.S. lessons; in fact, only two of the nine U.S. teachers who use manipulatives to model the process of finding a mean make a connection between the evening-out process and the written form of the averaging algorithm. In addition to the use of manipulatives for providing students with accessible facilities to learn the concept of average, U.S. teachers also use manipulatives or physical activities to generate data. For example, students were asked to actually measure their heights and the lengths of their arms, and then see if "each student is average." There are not any Chinese lessons requiring students to generate their own data.

These finding suggests that U.S. teachers encourage students less frequently at this grade level to move to more abstract, conventional representations and strategies in their classroom instruction. One of the common misconceptions held by many U.S. teachers is that concrete representations or manipulatives are the basis for all learning since they believe that concrete representations or manipulatives can facilitate students' conceptual understanding (Burrill, 1997). However, research has shown that manipulatives or concrete representations do not guarantee students' conceptual understanding (e.g., Baroody, 1990). The purpose of using concrete, visual representations is to mediate students' conceptual understanding of the abstract nature of mathematics, but concrete experiences do not automatically lead to generalization and conceptual understanding. Unfortunately, the U.S. teacher whose lessons we videotaped made no attempt to help students make the transition from concrete, visual representations to symbolic representations.

CHINESE AND U.S. INTENDED CURRICULAR TREATMENTS

While we have focused on different mathematical topics to examine Chinese and U.S. intended curricular treatments, in this section, I will highlight some findings related to the intended treatment of arithmetic average in two U.S. National Science Foundation-funded curricula and one Chinese curriculum (Division of Mathematics, 1996). The two U.S. NSF-funded curricula include *Mathematics in Context* (MiC) (National Center for Research in Mathematical Sciences Education at the University of Wisconsin/Madison and Freudenthal Institute at the University of Utrecht, 1997-1998) and the *Connected Mathematics Program* (CMP) (Lappan, Fey, Fitzgerald, & Friel, 1998).

It should be indicated that, unlike the Chinese curriculum, the two U.S. National Science Foundation-funded curricula consist of a number of unit booklets for each grade level. Therefore, for the two U.S. series, only the units with a focus on the concept of average were included in the study. In particular, two units from *Mathematics in Context* are included in this study: "Picturing Numbers" and "Dealing with Data." For *Connected Mathematics*, only the units called "Data About Us" and "Samples and Populations" were included.

INTRODUCTION OF THE CONCEPT

The Chinese curriculum uses an evening-out approach to illustrate the meaning of average and the processes of finding an average. Then students are guided to connect the even-out process with the "add-and-divide" algorithm of finding an average. In particular, the China series first uses "equal dividing" or "equal-sharing" to refer to the notion of average, and then introduces the even-out process to find the average of several numbers. The worked-out example that introduces the concept is related to evening out four cups of water: "*There are four cups of water. The water in the first cup is 6 cm high, 3 cm high in the second, 5 cm high in the third,*

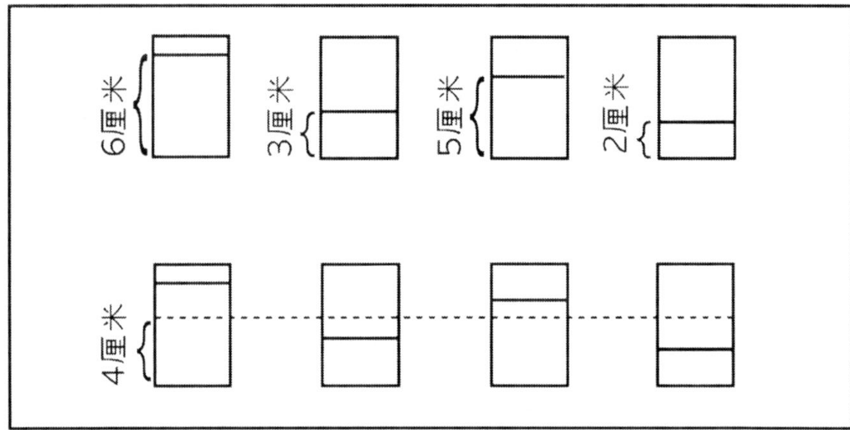

Figure 10.3.   A diagram showing evening-out process in the Chinese curriculum.

and 2 cm high in the fourth. What is the average height of the water level in the four cups?" The student textbook contains a diagram, shown in Figure 10.3 to even out the water levels in the four cups in order to find the average. Then the evening-out process is related to the algorithmic approach as $(6 + 3 + 5 + 2)/4 = 16/4 = 4$ cm. Based on this worked-out example, the formula for finding an average is formally introduced: average = total /number of items. The Chinese curriculum did not introduce the concept as the representative of a data set.

The introduction in the two U.S. series emphasizes students' understanding of the concept as the representative of a data set rather than showing how to actually calculate the arithmetic average. For example, in *Mathematics in Context*, the concept is first introduced as a one-number summary to describe a data set. A graphic approach is used to help students understand the representativeness of the average. The teachers' manual explicitly indicates that students do not need to learn how to compute the average at this point (5th grade). Like *Mathematics in Context*, *Connected Mathematics* places more emphasis on the students' understanding of the concept as the representative of a data set rather than as the algorithm to calculate an average. However, unlike *Mathematics in Context*, *Connected Mathematics* uses an approach similar to the evening-out approach to open the discussion of the concept of average through stacking cubes and stick-on notes. It then focuses on using the arithmetic average to help describe a set of data and compare data sets. The evening-out process is modeled by using cubes and stick-on notes. The teachers' manual clearly states that the arithmetic average is a kind of balance point in the distribution. The purpose of using these models is to support the development of the algorithm for finding the mean: adding up all the numbers and dividing by the total number of numbers. This emphasis is different from that of the Chinese curriculum where the evening-out process also highlights the meaning of the average.

ANALYSIS OF WORKED-OUT EXAMPLES AND PRACTICE PROBLEMS

Worked-out examples and practice problems can be classified into three categories to capture the kinds of understanding of the average concept that each curriculum promotes. Figure 10.4 describes these categories, and within each category, different types of problems are identified. It clearly shows that the two U.S. series focus more on the statistical aspect of the concept of the average (as a representative of a data set); the Chinese series focus more on the concept as an algorithm. In the Chinese series, all of the worked-out examples or practice problems in the fourth grade involve direct application of the averaging algorithm (A1 and A2), and the majority of the worked-out examples or practice problems in the fifth grade involve reverse or flexible application of the averaging algorithm (B1, B2, B3, and B4). In particular, the Chinese fifth grade textbook contains a considerable number of weighted mean problems in various contexts. Only several problems in the Chinese series require understanding the concept of the average as a representative of a set of data in the fifth grade textbook (C1).

A) Direct application of average algorithm:
   A1: Given data set represented in table or graph and find average. ***China, CMP, MiC***
   A2: Given average and number of quantities, and then to find the total. ***China, MiC***

B) Flexible application of average algorithm:
   B1: Given average and a data set with one or more missing numbers, find the missing numbers. ***China***
   B2: Create a data set with a given average and number of data. ***China, CMP, MiC***
   B3: Find an overall average from multiple data sets. ***China***
   B4: Weighted average problem. ***China***

C) Appropriate interpretation and use of mean in statistical context:
   C1: Using the average to compare two data sets with unequal items. ***China, CMP***
   C2: Mean does not have to equal to any data. ***CMP, MiC***
   C3: Using mean together with range will give a fuller picture of the data set. ***CMP, MiC***
   C4: Compare the mean, median and mode and decide the appropriate usage of them. ***CMP, MiC***

Figure 10.4. Types of worked-out examples.

Although these two U.S. series also contain many problems requiring direct application of averaging algorithm, students are only required to find an estimate for an average or to use blocks to find an average. In *Mathematics in Context*, the first problem involving the computation algorithm of the average is the following: *In order to find mean number of puppies ten mother dogs have, one took the total number of puppies and divided that number by the total number of mother dogs. There are 30 puppies and ten mother dogs, so the mean is 30/10 = 3. Do you think the strategy always works? Why or why not?* The two U.S. series also contain quite a few problems that require students to determine the appropriateness of the mean, median, and mode in the situations in order to solve the problems.

For the two U.S. reform textbook series, the major focus is on the appropriate use and interpretation of mean in statistical contexts (C1, C2, C3, and C4 for *Connected Mathematics*; C2, C3 and C4 for *Mathematics in Context*). In both series, the mean is introduced along with other measures of central tendency and statistical graphs such as bar graphs and histograms. The differences among the three measures of the central tendency and appropriateness of using these measures in problem situations are explicitly discussed in the two U.S. series.

The reason that the Chinese curriculum does not introduce the mean, median, and mode at the same time might be related to the fact that the Chinese curriculum first treats the average as a per-unit-quantity through equal-dividing, instead of as a measure of central tendency. However, the two U.S. series first treat the average from a statistical point of view. The two U.S. reform series embed the initial treatment of the concept of average in the context of discussing data analysis and graph interpretation. Unlike the U.S. series, the three Asian series separate the initial introduction of average concept from the general discussion of data analysis and graph interpretation, which appears in the later grades.

FINAL REMARKS

In this paper, I have highlighted some similarities and differences in intended, planned/implemented, and achieved curricula between China and the United States. Chinese and U.S. students not only performed differently on various tasks, but also used different solution representations and strategies. Chinese students are more likely than U.S. students to use conventional strategies and abstract representations, but U.S. students are more likely to use unconventional strategies and concrete representations. These differences between Chinese and U.S. students can be readily explained by Chinese and U.S. teachers' differential conceptions and representations, as well as by intended differing treatments of curricula in China and the United States.

Identifying similarities and differences between China and the United States is interesting and important, but it is not the ultimate goal in my research agenda. The ultimate goal is to use the findings to improve both Chinese and U.S. students' learning. I have started to discuss lessons we can learn from a cross-national project like this elsewhere (e.g., Cai, 2001; Cai & Cifarilli, 2004). While the focus of this paper is to highlight some similarities and differences between China and the United States in school mathematics, I sincerely hope that the findings highlighted here can serve as a springboard for readers to reflect on issues related to the teaching and learning of mathematics, and provide helpful insights when we strive for improving students' learning of mathematics.

ACKNOWLEDGMENT

Research reported here has been supported by grants from the National Academy of Education, National Science Foundation, and Spencer Foundation. However, any opinions expressed herein are those of the author and do not necessarily represent the views of any of the Foundations.

APPENDIX A. TASKS AND RELATED STUDENT RESPONSES

Hats Average Problem

Angela is selling hats for the Mathematics Club. This picture shows the number of hats Angela sold during the first 3 weeks.

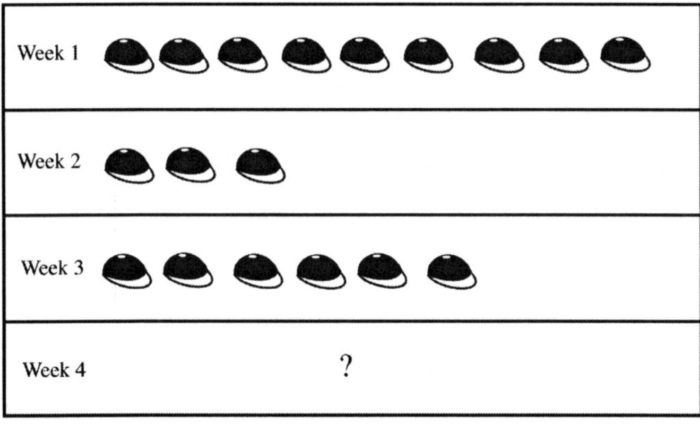

How many hats must Angela sell in Week 4 so that the *average* number of hats sold is 7? Show how you found your answer.

Response A

N = the number of hats sold in Week 4.

$9 + 3 + 6 + n = 4 \times 7$. $18 + n = 28$. $n = 28 - 18 = 10$. Angela sold 10 hats in Week 4.

Response B

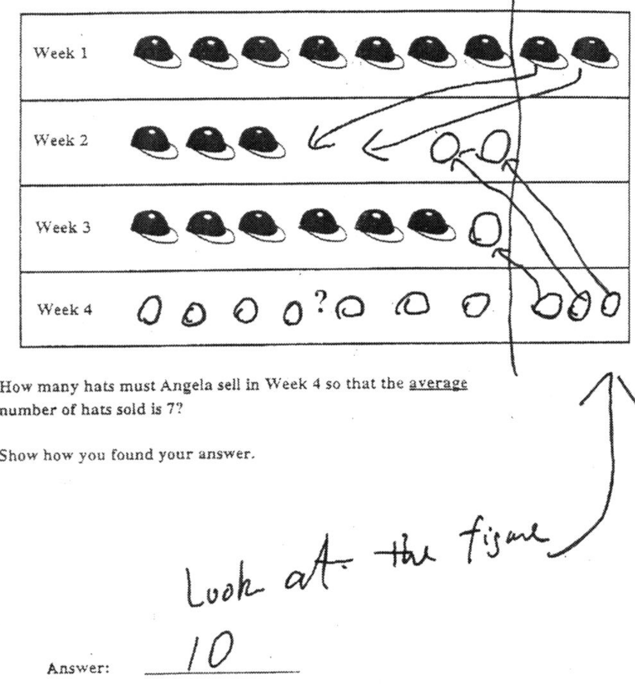

Response C

$9 + 3 + 6 = 18$. $4 \times 7 = 28$. $28 - 18 = 10$. So 10 hats were sold in Week 4.

Score Average Problem

The average of Ed's 10 test scores is 87. The teacher throws out the top and bottom scores, which are 55 and 95. What is the average of the remaining set of scores? Show how you found your answer.

Response D

10 × 87 = 870. 870 – 55 – 95 = 720. 720/8 = 90. The average of the remaining 8 scores is 90.

Response E

The student first used one of the properties of average and determined that the average for the remaining eight scores must be between 55 and 95. Then the student drew ten circles and put 95 in the first and 55 in the last, leaving eight empty circles. Using a modified sharing approach, the student realized that 55 and 95 contributed 15 to the average [(95 + 55) ÷ 10 = 15]. So the student said that each of the eight blank spaces should get 15. But 15 is 72 less than 87 (the average for the 10 scores), the student then multiplied 10 by 72 and got 720. 720 ÷ 8 = 90. Thus, 90 became the average of the remaining eight scores after the top and bottom scores were thrown away.

Response F

I think that the average for the remaining set of scores is between 55 and 95. But 87 is closer to 95 than 55. So the average for the remaining must be about 90.

Elevator Average Problem

There are 10 people in an elevator, 4 women and 6 men. The average weight of the women is 120, and the average weight of the men is 180 pounds. What is the average of the weights of the 10 people in the elevator? Show how you found your answer.

Cost Average Problem

The average of costs for 10 books is $24. What would be the cost for each of the 10 books? Show how you found your answer.

Odd Number Pattern Problem

Sally is having a party.
The first time the doorbell rings, 1 guest enters.
The second time the doorbell rings, 3 guests enter.

The third time the doorbell rings, 5 guests enter.

The fourth time the doorbell rings, 7 guests enter.

Keep going in the same way. On the next ring a group enters that has 2 more persons than the group that entered on the previous ring.

(a)   How many guests will enter on the 10th ring?
Explain or show how you found your answer.

(b)   In the space below, write a rule or describe in words how to find the number of guests that entered on each ring.

(c)   99 guests entered on one of the rings. What ring was it?

Explain or show how you found your answer.

Baby Average Problem

Which animal has more babies per litter: a mouse or a cat?

Grace and Huong wanted to find out. First, they collected as much data as possible. They asked all the people in their school who have mice or cats whether or not their pets have ever had a litter and, if so, how many babies were in the litter. Grace and Huong collected the following data about 18 mice and 12 cats.

Number of Babies in Litter (Mice)

Mouse	1	2	3	4	5	6	7	8	9
A	🐭	🐭	🐭						
B	🐭	🐭	🐭	🐭					
C	🐭	🐭	🐭	🐭	🐭	🐭	🐭	🐭	🐭
D	🐭	🐭	🐭						
E	🐭	🐭	🐭	🐭	🐭				
F	🐭	🐭							
G	🐭	🐭	🐭	🐭	🐭	🐭	🐭	🐭	🐭
H	🐭	🐭	🐭	🐭	🐭				
I	🐭	🐭	🐭						
J	🐭	🐭							
K	🐭	🐭	🐭	🐭					
L	🐭	🐭	🐭	🐭	🐭	🐭	🐭	🐭	🐭
M	🐭	🐭	🐭	🐭	🐭				
N	🐭	🐭							
O	🐭	🐭	🐭	🐭	🐭	🐭	🐭	🐭	🐭
P	🐭	🐭							
Q	🐭	🐭	🐭						
R	🐭	🐭	🐭	🐭	🐭				

Number of Babies in Litter (Cats)

Cat	1	2	3	4	5	6	7	8	9	10
A	🐱	🐱	🐱	🐱						
B	🐱	🐱	🐱	🐱	🐱	🐱				
C	🐱	🐱	🐱							
D	🐱	🐱	🐱	🐱	🐱					
E	🐱	🐱	🐱	🐱	🐱					
F	🐱	🐱	🐱							
G	🐱	🐱	🐱	🐱	🐱	🐱	🐱	🐱		
H	🐱	🐱	🐱							
I	🐱	🐱	🐱	🐱	🐱	🐱	🐱			
J	🐱	🐱	🐱	🐱	🐱					
K	🐱	🐱	🐱	🐱						
L	🐱	🐱	🐱	🐱	🐱	🐱	🐱	🐱	🐱	🐱

The Pizza Ratio Problem

Here are some children and pizzas. 7 girls share 2 pizzas equally and 3 boys share 1 pizza equally.

(a) Does each girl get the same amount as each boy?
    Explain or show how you found your answer.
(b) If each girl does not get the same amount as each boy, who gets more?
    Explain or show how you found your answer.

REFERENCES

Baroody, A. J. (1990). How and when should place-value concepts and skills be taught? *Journal for Research in Mathematics Education, 21,* 281–286.

Bruner, J. (1996). *The culture of education.* Cambridge, MA: Harvard University Press.

Burrill, G. (1997). The NCTM standards: Eight years later. *School Science and Mathematics, 97*(6), 335–339.

Cai, J. (1995). A cognitive analysis of U.S. and Chinese students' mathematical performance on tasks involving computation, simple problem solving, and complex problem solving. *Journal for Research in Mathematics Education, Monograph series, Vol. 7.*

Cai, J. (1997). Beyond computation and correctness: Contributions of open-ended tasks in examining U.S. and Chinese students' mathematical performance. *Educational Measurement: Issues and Practice, 16*(1), 5–11.

Cai, J. (2000a). Mathematical thinking involved in U.S. and Chinese students' solving process-constrained and process-open problems. *Mathematical Thinking and Learning, 2,* 309–340.

Cai, J. (2000b). Understanding and representing the arithmetic averaging algorithm: An analysis and comparison of U.S. and Chinese students' responses. *International Journal of Mathematical Education in Science and Technology, 31*(6), 839–855.

Cai, J. (2001). Improving mathematics learning: Lessons from cross-national studies of U.S. and Chinese students. *Phi Delta Kappan, 82*(5), 400–405.

Cai, J. (2003). Investigating parental roles in students' learning of mathematics from a cross-national perspective. *Mathematics Education Research Journal, 15*(2), 87–106.

Cai, J. (2004a). Why do U.S. and Chinese students think differently in mathematical problem solving? Exploring the impact of early algebra learning and teachers' beliefs. *Journal of Mathematical Behavior, 23,* 135–167.

Cai, J. (Ed.). (2004b). Developing algebraic thinking in the earlier grades: Case studies of the Chinese, Russian, Singaporean, South Korean, and U.S. school mathematics (Special issue). *The Mathematics Educators, 8*(1).

Cai, J. (2005). U.S. and Chinese teachers' knowing, evaluating, and constructing representations in mathematics instruction. *Mathematical Thinking and Learning: An International Journal, 7*(2), 135–169.

Cai, J. (2006). U.S. and Chinese teachers' cultural values of representations in mathematics education. In F. K. S. Leung, K. D. Graf, & F. Lopez-Real (Eds.), *Mathematics education in different cultural traditions: A comparative study of East Asian and the West* (pp. 465–482). New York: Springer.

Cai, J., & Cifarelli, V. (2004). Thinking mathematically by Chinese learners: An international comparative perspective. In L. Fan, N. -Y. Wong, J. Cai, & S. Li (Eds.), *How Chinese learn mathematics: Perspectives from insiders* (pp. 71–106). Singapore: World Scientific Publishers.

Cai, J., & Hwang, S. (2002). U.S. and Chinese students' generalized and generative thinking in mathematical problem solving and problem posing. *Journal of Mathematicsthematical Behavior,* 21(4), 401–421.

Cai, J., & Lester, F. K., Jr. (2005). Solution and pedagogical representations in Chinese and U.S. mathematics classroom. *Journal of Mathematical Behavior, 24*(3-4), 221–237.

Cai, J., & Sun, W. (2002). Developing students' proportional reasoning: A Chinese perspective. In B. H. Litwiller & G. W. Bright (Eds.), *Making sense of fractions, ratios and proportions: 2002 yearbook* (pp. 195–205). Reston, VA: National Council of Teachers of Mathematics.

Cai, J., & Wang, T. (2006). U.S. and Chinese teachers' conceptions and constructions of representations: A case of teaching ratio concept. *International Journal of Mathematics and Science Education, 4*(1), 145–186.

Cai, J., Lo, J. J., & Watanabe, T. (2002). Intended treatments of arithmetic average in U.S. and Asian school mathematics. *School Science and Mathematics, 102*(8), 391–404.

Cai, J., Lew, H. C., Morris, A., Moyer, J. C., Ng, S. F., & Schmittau, J. (2005). The development of students' algebraic thinking in earlier grades: A cross-cultural comparative perspective. *Zentralblatt fuer Didaktik der Mathematik* [International Review on Mathematics Education], *37*(1), 5–15.

Darling-Hammond, L. (1994). Performance-based assessment and educational equity. *Harvard Education Review, 64*(1): 5–30.

Division of Mathematics. (1996). *National unified mathematics textbooks in elementary school*. Beijing, China: People's Education Press.

Fennema, E., & Franke, M. L. (1992). Teachers' knowledge and its impact. In D. A. Grouws (Ed.), *Handbook of research on mathematics teaching and learning* (pp. 147–164). New York: Macmillan.

Gardner, H. (1989). *To open minds: Chinese clues to the dilemma of contemporary education*. New York: Basic Books.

Gardner, H. (1999). *The disciplined mind: What all students should understand*. New York: Simon & Schuster.

Lappan, G., Fey, J. T., Fitzgerald, W. M., & Friel, S. N. (1998). *Connected mathematics*. Menlo Park, CA: Dale Seymour Publications.

Ma, L. (1999). *Knowing and teaching elementary mathematics: Teachers' understanding of fundamental mathematics in China and the United States*. Hillsdale, NJ: Erlbaum.

National Center for Research in Mathematical Sciences Education at the University of Wisconsin/Madison and Freudenthal Institute at the University of Utrecht. (1997–1998). *Mathematics in context*. Chicago: Britannica.

Stigler, J. W., & Hiebert, J. (1999). *The teaching gap: Best ideas from the world's teachers for improving education in the classroom*. New York: The Free Press.

Wittrock, M. C. (Ed.). (1986). *Handbook of research on teaching* (3rd ed.). New York: Macmillan.

CHAPTER 11

TRANSFORMING CURRICULUM FROM INTENDED TO IMPLEMENTED

What Teachers Need to Do and What They Learned in the United States and China

Yeping Li
Texas A&M University

INTRODUCTION

Cross-system studies of mathematics curricula and teacher education provide unique opportunities to understand the differences and similarities among different systems, and suggest alternative ways for future teacher training for curriculum transformation and implementation. In this paper, I first summarize findings from some relevant studies probing how U.S. and Chinese mathematics textbooks are structured for teaching and learning. Cross-system variations in textbooks present teachers with different opportunities and challenges for classroom instruction. I then present collaborative research efforts investigating approaches and expec-

Mathematics Curriculum in Pacific Rim Coutries—China, Japan, Korea, and Singapore:
Proceedings of a Conference, pp. 183–195

tations in preparing teachers for classroom instruction in the United States and China. The results show different approaches and emphases in teacher preparation in terms of what prospective teachers need to learn about curriculum and instruction. Finally, possible connections and disconnections between teacher training and what teachers need to do in curriculum transformation, and implications obtained from cross-system comparisons are discussed.

Teaching and learning activities in modern schools are organized in specific ways. The requirements in both subject content and performance are given in a way that structures students' learning experiences in schools. Such requirements in content and performance are more or less specified in the form of school curriculum (Schmidt McKnight, Valverde, Houant, & Wiley, 1997). Studies of school curriculum, therefore, can inform the researchers about the expectations, processes, and outcomes of students' learning experiences in schools. Cross-system studies of mathematics curriculum are thus important for the design and interpretation of a cross-system comparative study of students' mathematics achievement, and for the improvement of mathematics curriculum and its implementation in different systems.

Because curriculum materials have been a mainstay in mathematics classrooms in many education systems (Howson, 1995; McKnight et al., 1987; Schmidt et al., 1997), it is reasonable to assume that curriculum materials provide more or less detailed guidelines for daily teaching and learning activities in classrooms. In fact, advocates of school mathematics reform in the United States tend to influence classroom practice, and hence, student attainment, by means of making changes in curriculum materials (Begle, 1973; Senk & Thompson, 2003). Curriculum materials, as designed and produced, often contain different components and are structured in different ways. These differences afford varied opportunities for teaching and learning. An examination of curriculum materials themselves is thus important to reveal possible opportunities and challenges presented to teachers for developing their classroom instruction.

Although curricular materials play an important role in shaping what is taught and learned in classrooms (Schmidt et al., 1997; Schmidt, McKnight, Houang, Wong, & Wiley, 2001), the curriculum enacted in classrooms often combines with teachers' own thinking and planning (Doyle, 1993; Remillard, 1999). The design and variations of curriculum materials thus present only one part of the story of what is taught and learned in classroom settings. Possible influences of curriculum materials on classroom instruction are relevant to what teachers know and do in transforming curriculum from the intended, such as curriculum framework and textbooks, to what is implemented in classrooms. This paper is therefore structured to address the following two questions:

- What opportunities and challenges may different textbooks present to mathematics teachers for developing their classroom instruction?
- What have teachers learned (or been prepared) to transform curriculum from the intended to the implemented?

To address these two questions, I will bring together two lines of study in an international context:

1. Cross-system studies of mathematics textbooks that reveal different opportunities and challenges for teaching and learning mathematics;
2. Studies on mathematics teacher preparation that reveal what pre-service mathematics teachers are taught about curriculum and instruction in methods courses.

In particular, I will focus on some relevant studies related to the United States and China. It is assumed that differences between these two educational systems will help bring the needed contrast that otherwise is not available within either single system. The following figure depicts some general differences between these two systems. As shown in Figure 11.1, what is taught in the classroom bears influences from several different sources in the United States, but not in China. Textbooks dominate the influence on the implemented curriculum in the case of China but not in the United States. Thus, teachers in the United States and China face different challenges in determining, structuring, and organizing what is to be taught in classrooms.

CROSS-SYSTEM DIFFERENCES IN OPPORTUNITIES AND CHALLENGES EMBEDDED IN MATHEMATICS TEXTBOOKS FOR TEACHING

Mathematics textbooks have typically been taken as providing specifications about content topic sequence, expectations, content presentation and organization (e.g., Valverde, Bianchi, Wolfe, Schmidt, & Houang, 2002). They are used as a content resource for both teachers and students in the process of teaching and learning mathematics, also as a guide for structuring teaching and learning activities. Relevant efforts to examine mathematics textbooks have led to different academic interests that include: (1) content topics, (2) students' performance expectations, and (3) content presentation and organization features.

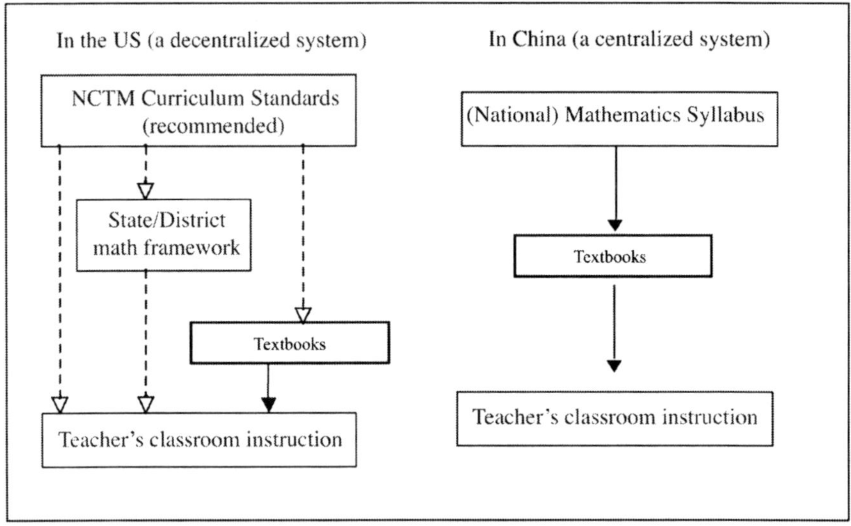

Note: The arrow with a solid line shows a direct and substantial influence, the arrow with a dashed line shows a general influence that can vary substantially from case to case.

Figure 11.1.   System context differences reflected in structuring what is taught in classrooms from a curriculum point of view.

Cross-system studies on content topics in mathematics textbooks include the examination of (a) content topic inclusion at a specific grade level, (b) grade-level placement of specific content topics, (c) content topic pace and emphasis, (d) inclusion of new content topics across grade levels, and (e) content treatment of a specific topic (e.g., Cai, Lo, & Watanabe, 2002; Fuson, Stigler, & Bartsch, 1988; Li, 1999, 2002; Schmidt et al., 1997). The examination of content topic inclusion in mathematics textbooks helps reveal possible opportunities provided by textbooks for student learning. The results from some comparative studies between American and Chinese mathematics textbooks indicate a pattern of consistent differences in selected elementary and middle school mathematics textbooks (e.g., Fuson et al., 1988; Li, 1999, 2002; Schmidt et al., 1997). In general, American textbooks contain many more content topics, more mathematical connections with other scientific disciplines, and sequence topics at a slower pace and with a spiral approach. In contrast, Chinese textbooks include fewer but mathematically challenging content topics, less mathematical connections with other scientific disciplines, and sequence topics at a fast pace and with a linear approach.

**Table 11.1.   Cross-System Differences in Selected
Textbook Features Between the U.S. and China**

U.S. Mathematics Textbooks	*Chinese Mathematics Textbooks*
Contain many different content topics and slow pace of topic sequence	Contain fewer but mathematically challenging content
Contain many small and diverse units	Show a fast pace of topic sequence
Show a step-by-step introduction of mathematics concepts, including many illustrations and exercise problems	Contain large units with a concentration on content introduction
Emphasize connections with science, real world, and others	Emphasize content explanation and using worked-out examples
Contain many more and diverse exercise problems for developing students cognition in general and problem-solving ability	Contain limited number and types of exercise problems mainly for facilitating students' acquisition of newly introduced mathematics concepts or procedures

Likewise, existing cross-national studies of American and Chinese mathematics textbooks also examined expectations for student performance (e.g., Li, 1999, 2000; Schmidt et al., 1997; Stigler, Fuson, Ham, & Kim, 1986) and features for teaching organization and learning activities (e.g., Carter, Li, & Ferrucci, 1997; Li, 1999). A consistent pattern of cross-system differences is summarized in Table 11.1.

The differences between American and Chinese mathematics textbooks, as summarized above, suggest different opportunities that U.S. and Chinese textbooks provide to teachers. Table 11.2 contains a summary of these opportunities. The detailed design and broad inclusion of content topics in the U.S. textbooks present various topic choices and a user-friendly format of content organization for teachers. In contrast, the concise presentation and coherent organization of content topics in Chinese textbooks provide room for teachers to expand and reorganize content for classroom instruction.

At the same time, the United States and China have very different education systems. When textbooks are used for classroom instruction, cross-system differences in textbooks pose different kinds of challenges for teachers (see Table 11.2). While Chinese teachers need to focus on expanding and reorganizing *content* from what is presented in textbooks, U.S. teachers need to take responsibility to reorganize *curriculum* for classroom instruction. Therefore, the use of American textbooks require the teacher to be a specialist in curriculum at both macro and micro levels for planning and structuring what is to be taught in classrooms. The macro level refers to the large scope of curriculum that teachers need to align

Table 11.2.   Cross-System Differences in Opportunities and Challenges Provided by Textbooks Between the U.S. and China

	Opportunities	*Challenges*
U.S. mathematics textbooks	• Present many possible choices of content topics, units, and exercise problems • Almost ready-to-be-used lesson structure • Mathematics content is presented as small steps and fun through connecting mathematics to real world, etc.	• Teachers need to decide how to align state or school district mathematics framework with textbooks in terms of both content and students' performance requirements • Teachers need to decide how to select and/or reorganize different units. • Teaching and learning mathematics through problem solving is an advocated approach but is more difficult to implement than we think
Chinese mathematics textbooks	• Present and organize coherent mathematics content in large units • Leave rooms for content expansion • Mathematics content is presented in a way that is easy for students to read and review	• Expect further efforts from teachers to develop and structure lessons • Expect further efforts from teachers to expand content and/or select other problems to enrich classroom instruction

between what is required by a state's and a school district's mathematics framework and what can be found in textbooks. The micro level refers to the small scope of curriculum (e.g., lessons) that teachers need to plan and structure. Moreover, U.S. teachers need to be specialists in instruction for considering the pedagogical aspect of curriculum planning and for enacting curriculum in classrooms. However, the content inclusion and design of U.S. mathematics textbooks place no special expectations for teachers at elementary and middle school levels to be content specialists in mathematics.

In contrast, Chinese mathematics textbooks require teachers to be specialists in content, curriculum (mainly at the micro level), and instruction. The concise design of textbooks and the inclusion of mathematically challenging content require further efforts from teachers in planning and organizing content for classroom instruction. However, the centralized education system in China places no specific requirements on teachers to align textbooks and curriculum framework. Therefore, while the U.S. teachers need to know more about curriculum alignment and transformation but not necessarily mathematics content, Chinese teachers are in an opposite position. Since the teacher is key to effective classroom instruction (e.g., National Council of Teachers of Mathematics, 1996), cross-system differences in requirements placed on teachers suggest the importance of examining what teachers have learned before entering their profession.

PROSPECTIVE TEACHER PREPARATION
FOR TEACHING MATHEMATICS

Across China and the United States, helping preservice teachers learn to teach is an important component of teacher education programs in both systems. However, cross-national comparisons of what preservice teachers are expected to learn in teacher education programs have not been studied extensively (Li & Lappan, 2003). Some existing studies examined what preservice teachers are expected to learn about curriculum and instruction in the "methods" courses in the United States and China, respectively (e.g., Li, 2003; Philipp, Thanheiser, & Clement, 2003). Here I try to pull several studies together to highlight some cross-system similarities and differences on the approaches and expectations for what preservice teachers need to learn about curriculum and instruction in the "methods" courses.

In the United States, a methods course that is normally called such has a focus on helping preservice teachers learn various instructional strategies. For example, at the Mathematics Department of University of New Hampshire (UNH), there is a methods course designed for preservice elementary school teachers—MATH 703: Teaching of Mathematics, K-6. This is a 4-credit hour course with a focus on methods of teaching. Many teacher education programs in the United States seem to share quite a few similarities to this course. For example, Graham, Li, and Buck (2000) conducted a survey of selected mathematics teacher preparation programs in the United States, and found program characteristics in the sample similar to what had been in place for the past 75 years. The standards-based reform movement did not bring substantial changes to teacher preparation programs, except that the NCTM standards document (2000) has often been used in the methods course. Also, the teaching of methods courses in many teacher education programs in the U.S. often incorporates children's thinking into the course content (Philipp et al. 2003). Building on what students know and can learn for teaching is a broadly accepted philosophy and is highlighted in helping preservice teachers learn how to teach. Some educators, like Philipp and his colleagues (2003), go even further to implement and study a model to integrate mathematics content and children's mathematical thinking at a very early stage in the teacher education program. It indicates an ever increasing trend in teaching methods courses in the United States: prospective teachers need to learn how to help students construct mathematics knowledge based on what children know and can do.

In summary, methods courses for preservice teachers in the United States bear some general features in common. First, its content focus is instructional methods. Second, no textbook is designated for the methods course. Instructors can make or select one as they like, and NCTM *Principles and Standards for School Mathematics* (2000) is often used as a textbook in the methods course. Finally, diverse approaches have been used to teach the methods course, but there is one shared tendency of helping prospective teachers learn how to teach based on how students may learn mathematics.

In China, the methods course is called "teaching materials and methods" as translated literally word by word. Because China has a centralized education system, it has a unified curricular guide for both school students' learning and teacher preparation (Sun, 2000). For the elementary mathematics methods courses in China, its content is structured as containing two parts: (1) fundamental mathematics that pre-service teachers are going to teach in elementary school classrooms (1st methods course); and (2) teaching materials and methods (2nd methods course). The textbooks used for the methods courses are "elementary school mathematics teaching materials and methods" plus elementary school mathematics textbooks. The instructional approach used to teach these methods courses is dominated by lectures although there are other activities. The 2nd methods course emphasizes knowing, understanding, and exploring the content and formation of mathematics textbooks. In contrast to the methods courses in the U.S., the methods course in the Chinese teacher education program has a special feature: lecturing, exploring, and making school mathematics curriculum as a way of helping prospective elementary school teachers learn how to teach (Li, 2003). In particular, elementary school mathematics textbooks are used as a regular part of the methods course instruction; emphasis is placed on the introduction of national mathematics syllabus, deeper understanding of elementary school mathematics content, general pedagogy and specific methods for organizing elementary mathematics content for teaching; and the course instruction follows the course textbook closely.

Comparing approaches and expectations in preparing prospective teachers for classroom instruction in the U.S. and China, different focuses emerge. In the United States, preservice mathematics teachers learn to become specialists in instruction (with a strong tendency towards knowing about student learning) and curriculum (mostly at the micro level). In

China, however, pre-service mathematics teachers learn to become specialists in content, curriculum, and instruction.

WHAT TEACHERS NEED TO DO VERSUS
WHAT THEY LEARNED ABOUT WHAT TO DO

As discussed previously, effective use of mathematics textbooks poses specific requirements on teachers. In the case of using U.S. textbooks, U.S. teachers need to be specialists in *curriculum* (at the macro and micro levels) and instruction, but not necessarily specialists in content at elementary and middle school levels. Through taking methods courses during their teacher education program, however, U.S. preservice mathematics teachers learn to become specialists in *instruction* and curriculum (most likely at the micro level only). Although there are some connections between what teachers are required to do and what they learned about what to do, U.S. teachers lack some important training that is needed by the profession. In particular, the context of the system puts high demands on teachers in aligning what is to be taught in classrooms with intended curriculum as given by state and school district curriculum frameworks and textbooks. However, U.S. teachers learn more about pedagogy and student learning in their teacher education programs but not on curriculum alignment and transformation.

The discrepancy in what U.S. teachers would need to do with curriculum and what they learn is less obvious than the need for teachers to have more training in mathematics content. Many researchers have argued that content knowledge is an important component of teacher's expertise (e.g., Hill, Rowan, & Ball, 2005; Leinhardt & Smith, 1985; Ma, 1999) and have documented the need to enhance in-service and preservice teachers' content knowledge and pedagogical content knowledge in mathematics (e.g., Ball, 1990; Hiebert, Gallimore, & Stigler 2002). The call for teacher's knowledge improvement is clearly stemmed from research but not necessarily from reality. As revealed from the above discussions, in fact, the use of U.S. mathematics textbooks at the elementary and middle school levels does not really pose special challenges for teachers to be content specialists. In other words, without being content specialists, U.S. teachers can teach mathematics fairly easily with textbooks at the elementary and middle school levels.

In contrast, effective use of Chinese mathematics textbooks poses different challenges to teachers. On the one hand, Chinese teachers need to be specialists in content, curriculum (mainly at the micro level) and instruction to teach with Chinese mathematics textbooks. On the other hand, through taking methods courses in teacher education programs,

Chinese preservice mathematics teachers learn to become specialists in content, curriculum, and instruction. A strong and consistent connection can be seen between what is expected from teachers and what preservice teachers can learn from teacher education programs.

CONCLUDING THOUGHTS

This cross-system contrast between the United States and China suggests that U.S. teachers face different challenges in carrying out effective classroom instruction in comparison to Chinese teachers. First, the U.S. education system presents teachers with substantially more challenges in curriculum alignment and transformation than the case in China. While Chinese teachers only need to focus on curriculum transformation from textbooks to classroom instruction (at the micro level), U.S. teachers need to coordinate and align more curriculum information to plan curriculum for classroom instruction (at both the macro and the micro levels). However, Chinese teachers, but not U.S. teachers, receive quite substantial training and information about curriculum transformation and planning in their teacher education programs. Thus, many more programs and other types of helpful resources should be made available in the U.S. for teachers in aligning curriculum and developing challenging curriculum for classroom instruction.

Second, although teacher's content knowledge is broadly accepted as an important component of teacher's expertise, the design and use of textbooks present different challenges cross-nationally for teachers. In the case of the U.S., textbooks are well designed in a way that makes it possible for teachers to use them without being content specialists. The design and content requirement of Chinese mathematics textbooks, however, requires teachers to know more and more deeply than what is presented in textbooks. Chinese teachers are expected to be content specialists, an expectation that is fortunately supported by their preservice education. The cross-system differences suggest that U.S. teachers may lack the challenge and motivation to deepen their own content knowledge for teaching mathematics at the elementary and middle school levels. This reality is further "supported" by the training that preservice teachers receive from many teacher education programs in the United States. Therefore, an improvement of teacher's content knowledge can be theoretically sound, but making a change would take much more than what we can learn from research. It becomes necessary to create a working environment that would require teachers to have and use in-depth mathematics knowledge.

As teaching is a cultural activity (Stigler & Hiebert, 1999), the same holds true for mathematics curriculum and its transformation in school education. Cross-system differences in curriculum and teacher education may not be translated directly into suggestions for change. However, a cross-system comparison in textbooks and teacher education between the United States and China, as discussed above, shows that the U.S. teachers are required to assume more responsibilities in aligning and transforming curriculum for classroom instruction. They are also expected to improve their own content knowledge for designing and implementing a challenging curriculum. Since teachers are key to effective classroom instruction (e.g., National Commission on Teaching and America's Future, 1996), I would argue that more resources and supporting programs should be identified and developed in the United States, so that teachers needs are met thus allowing them to fulfill the specific responsibilities and expectations placed on them.

REFERENCES

Ball, D. L. (1990). Prospective elementary and secondary teachers' understanding of division. *Journal for Research in Mathematics Education, 21*(2), 132–144.

Begle, E. G. (1973). Lessons learned from SMSG. *Mathematics Teacher, 66,* 207–214.

Cai, J., Lo, J. J., & Watanabe, T. (2002). Intended treatments of arithmetic average in U.S. and Asian school mathematics. *School Science and Mathematics, 102*(8), 391–404.

Carter, J., Li, Y., & Ferrucci, B. (1997). A comparison of how textbooks present integer addition and subtraction in PRC and USA. *The Mathematics Educator, 2*(2), 197–209.

Doyle, W. (1993). Constructing curriculum in the classroom. In F. K. Oser, A. Dick, & J. Party (Eds.), *Effective and responsible teaching* (pp. 66–79). San Francisco: Jossey-Bass.

Fuson, K., Stigler, J., & Bartsch, K. (1988). Brief report: Grade placement of addition and subtraction topics in Japan, Mainland China, the Soviet Union, Taiwan, and the United States. *Journal for Research in Mathematics Education, 19*(5), 449–456.

Graham, K. J., Li, Y., & Buck, J. C. (2000). Characteristics of mathematics teacher preparation programs in the United States: An exploratory study. *The Mathematics Educator, 5*(1/2), 5-31.

Hiebert, J., Gallimore, R., & Stigler, J. (2002). A knowledge base for the teaching profession: What would it look like and how can we get one? *Educational Researcher, 31*(5), 3–15.

Hill, H. C., Rowan, B., & Ball, D. L. (2005). Effects of teachers' mathematical knowledge for teaching on student achievement. *American Educational Research Journal, 42*(2), 371–406.

Howson, G. (1995). *Mathematics textbooks: A comparative study of grade 8 texts*. Vancouver, British Columbia, Canada: Pacific Educational Press.

Leinhardt, G., and D. A. Smith. (1985). Expertise in mathematics instruction: Subject matter knowledge. *Journal of Educational Psychology, 77*(3), 247–271.

Li, Y. (1999). *An analysis of algebra content, content organization and presentation, and to-be-solved problems in eighth-grade mathematics textbooks from Hong Kong, Mainland China, Singapore, and the United States*. Unpublished doctoral dissertation, University of Pittsburgh, Pittsburgh, PA.

Li, Y. (2000). A comparison of problems that follow selected content presentations in American and Chinese mathematics textbooks. *Journal for Research in Mathematics Education, 31*(2), 234–241.

Li, Y. (2002). A comparison of content treatment for the teaching of percent in selected American and Chinese mathematics textbooks. In D. S. Mewborn, P. Sztajn, D. Y. White, H. G. Wiegel, R. L. Bryant, & K. Nooney (Eds.), *Proceedings of the twenty-fourth annual meeting of the North American Chapter of the International Group for the Psychology of Mathematics Education* (Vol. 4, pp. 1666–1669). Columbus, OH: ERIC Clearinghouse for Science, Mathematics, and Environment Education.

Li, Y. (2003). Knowing, understanding, and exploring the content and formation of curriculum materials: A Chinese approach to empower prospective elementary school teachers pedagogically. *International Journal of Educational Research, 37*(2), 179–193.

Li, Y., & Lappan, G. (Eds.). (2003). Developing and improving mathematics teachers' competence: Practices and approaches across educational systems. *International Journal of Educational Research, 37*(2).

Ma, L. (1999). *Knowing and teaching elementary mathematics: Teachers' understanding of mathematics in China and the United States*. Mahwah, NJ: Erlbaum.

McKnight, C. C., Crosswhite, F. J., Dossey, J. A., Kifer, E., Swafford, J. O., Travers, K. J., & Cooney, T. J. (1987). *The underachieving curriculum: Assessing U.S. school mathematics from an international perspective*. Champaign: Stipes.

National Commission on Teaching and America's Future. (1996). *What matters most: Teaching for America's future*. New York: Teachers College Press.

National Council of Teachers of Mathematics. (2000). *Principles and standards of school mathematics* [PSSM]. Reston, VA: NCTM.

Philipp, R. A., Thanheiser, E., & Clement, L. (2003). The role of a children's mathematical thinking experience in the preparation of prospective elementary school teachers. *International Journal of Educational Research, 37*(2), 195–210.

Remillard, J. T. (1999). Curriculum materials in mathematics education reform: A framework for examining teachers' curriculum development. *Curriculum Inquiry, 29*(3), 315–342.

Schmidt, W. H., McKnight, C. E., Valverde, G. A., Houang, R. T., & Wiley, D. E. (1997). *Many visions, many aims (Vol. 1): A cross-national investigation of curricular intentions in school mathematics*. Dordrecht, Netherlands: Kluwer Academic Press.

Schmidt, W. H., McKnight, C. C., Houang, R. T., Wang, H. C., Wiley, D. E., Cogan, L. S., et al. (2001). *Why schools matter—A cross-national comparison of curriculum and learning*. San Francisco: Jossey-Bass.

Senk, S. L., & Thompson, D. R. (2003). *Standards-based school mathematics curricula: What are they? What do students learn?* Mahwah, NJ: Erlbaum.

Stigler, J. W., Fuson, K. C., Ham, M., & Kim, M. S. (1986). An analysis of addition and subtraction word problems in American and Soviet elementary mathematics textbooks. *Cognition and Instruction, 3*(3), 153–171.

Stigler, J. W., & Hiebert, J. (1999). *The teaching gap: Best ideas from the world's teachers for improving education in the classroom*. New York: The Free Press.

Sun, W. (2000). Mathematics curriculum for preservice elementary teachers: The People's Republic of China. *Journal of Mathematics Teacher Education, 3*, 191–199.

Valverde, G. A., Bianchi, L. J., Wolfe, R. G., Schmidt, W. H., & Houang, R. T. (2002). *According to the book—Using TIMSS to investigate the translation of policy into practice through the world of textbooks*. Dordrecht, Netherlands: Kluwer Academic.

CHAPTER 12

TRANSLATING ELEMENTARY SCHOOL MATHEMATICS CURRICULUM

Isn't School Mathematics Universal?

Tad Watanabe
The Pennsylvania State University

INTRODUCTION

Japanese students historically have performed well on various international studies, including the recent Trends in International Mathematics and Science Study (TIMSS) and Program for International Student Assessment (PISA) studies. A persistent question that remains unanswered is what are the contributing factors to such high performance? The 1995 TIMSS also included a video study component that provided a glimpse into mathematics classrooms in Japan, Germany, and the United States. What we saw on the public release videos were three very different approaches to mathematics teaching. From this limited study, some came away with the impression that there was one right way to teach. However,

Mathematics Curriculum in Pacific Rim Coutries—China, Japan, Korea, and Singapore: Proceedings of a Conference, pp. 197–207

the 1999 TIMSS video study, which included other high achieving countries, showed that teaching styles among the high achieving countries also varied (Hiebert et al., 2003).

Another potential contributing factor for high achievement suggested by some researchers is the curriculum (e.g., Schmidt, McKnight, & Raizen, 1996). For example, Schmidt, Houang, and Cogan (2002) argued that the curricula of highest achieving countries seem to be more cohesive, coherent, and focused. By contrast, the U.S. curricula tended to be unfocused and repetitive, indicating limited expectation of mastery. The results of my earlier content analysis of the Japanese elementary mathematics curriculum materials are consistent with this perspective (Watanabe, 2007).

However, in this paper, I would like to discuss the Japanese elementary school curriculum from a slightly different angle. Recently, I had the opportunity to participate in the translation of Japanese curriculum materials into English. In one project, we translated the most widely used elementary school mathematics textbook series (Grades 1 through 6) in Japan. In the other project, we translated the teaching guide for Grades 1-6, a document published by the Japanese Ministry of Education, Culture, Sports, Science and Technology, that elaborates and explains the national course of study (COS) (Takahashi, Watanabe, & Yoshida, 2004). The lower secondary (Grades 7-9) school document is currently being translated. Participation in these translation projects provided a unique opportunity to understand the Japanese elementary school mathematics in depth, as we read every word on every page in these documents.

TRANSLATING JAPANESE CURRICULUM MATERIALS

Everyone knows that no two languages are completely parallel, with word for word correspondence. In other words, there is no "function" that translates one language into another. This difference between languages is often an indicator of what is significant for a particular culture. Many people have heard that the Eskimo language includes many more words for snow than does the English language.

As we began the project of translating the Japanese curriculum materials, we anticipated that there might be some words and phrases that are difficult to translate from Japanese into English because of the cultural differences. However, what we did not anticipate was that there might be some terms/phrases that are so important in mathematics teaching and learning that are also problematic when attempts are made to translate them into English. After all, we naively thought, school mathematics is school mathematics, isn't it? If language differences indicate what different cultures might perceive as significant, the difficulty we experienced while translating the curriculum materials might be indicators of different ideas Japan and the United States consider significant in school, in particular, elementary school mathematics. Although there were several terms and phrases that were difficult to translate, I want to focus on three words in this paper. Those words are *shiki* (式), *kufu* (工夫), and *sansu* (算数). In the next section, I will discuss what these words mean and share some examples illustrating the centrality of the ideas represented by these words in the Japanese elementary school mathematics.

SHIKI

This word is typically translated as "equation." In fact, the transcript of public release Japanese lesson 3 from the 1999 TIMSS Video Study included a phrase "inequality equation." This apparent contradictory phrase resulted due to the fact the Japanese term for inequality actually consists of three characters. Those characters mean, respectively, "not," "equal," and "*shiki*," thus, inequality equation. This example clearly shows that the Japanese word *shiki* actually describes something more than equations. In fact, in addition to equations and inequalities, *shiki* can be used to describe expressions such as $3 + 5$ or $x + 4$, as well as $3 + 5 =$. The project team finally decided to translate this word as "mathematical expression," but not "expression" in the technical sense. Rather, *shiki* encompasses many forms of expressions with mathematical symbols and notations that describe a quantity or a relationship among quantities.

Writing and interpreting *shiki* is one of the three main foci of the domain of Quantitative Relations in the Japanese elementary school mathematics curriculum. The learning expectations related to *shiki* in the 1989 elementary school mathematics COS are:

- Writing and interpreting mathematical expressions involving addition, subtraction, multiplication and division.
- Comparing numbers and quantities, and expressing their relationships.
- Writing and interpreting mathematical expressions involving ( ).
- Using $\square$, $\triangle$, a, and x in mathematical expressions, and evaluating such expressions by substituting specific values.

At the lower secondary level (Grades 7-9), *shiki* becomes a part of a domain, numbers and *shiki* (during his presentation at this conference, Mr. Yoshikawa translated the domain as "number and algebra," indicating, perhaps, *shiki* is more than just mathematical expressions). These

Figure 12.1.   A problem in a Grade 1 addition unit asks students to write *shiki* to describe the relationship among the quantities in the given pictures.

examples show the centrality of the notion of *shiki* in the Japanese school mathematics curriculum.

The treatment of *shiki* starts as soon as children start studying addition in the first grade. Figure 12.1 shows a problem at the beginning of the first addition unit in the translated Grade 1 textbook. As can be seen, the emphasis is on using *shiki* to describe the relationships among quantities. Moreover, students are not simply filling in the blanks with numbers, but they are expected to write a complete *shiki*, including the addition and equal signs.

Later in the year, the same textbook starts the unit on addition with sums greater than 10 with the following problem:

There are 9 persimmons on a tree and 4 persimmons on another tree. How many are there altogether?

(a) Please write a math sentence.
(b) Please think about how to do the calculation. (Sugiyama & Hironaka, 2000a, p. 62)

Here, students are first asked to describe the relationship among the quantities in a *shiki* to establish that the addition operation is indeed the appropriate operation in this situation. Only then are they asked to think about how to actually carry out the calculation.

These examples show that, even in Grade 1, *shiki* is treated as incorporating more than computation problems. This treatment is the way the authors of this textbook series try to address the following statements from the Japanese teaching guide:

> Teaching about writing and interpreting mathematical expressions already started at the stage of learning addition in first grade.... But, since concrete numbers are used in lower grades and calculation immediately leads to one number, children rarely become aware of the fact that 3 + 4 represents a concrete phenomenon. Therefore, it is important to teach children to focus on the meaning of mathematical expressions instead of paying attention solely to getting results. (Takahashi et al., 2004, p. 39)

The notion of "interpreting" *shiki* is often emphasized in a problem like the one shown in Figure 12.2. In this problem, children are asked to come up with a variety of ways in which the total number of dots in the arrangement can be calculated. One idea is presented as an example, accompanied by both an illustration and a mathematical expression. The textbook suggests that teachers can have students share their mathematical expressions while others try to determine their methods. Through this type of problem, textbooks attempt to help students understand that mathematical expressions are also useful in concisely representing and communicating their reasoning processes.[1]

In the intermediate grades, the Japanese textbook series include a discussion on how to combine multiple mathematical expressions into one

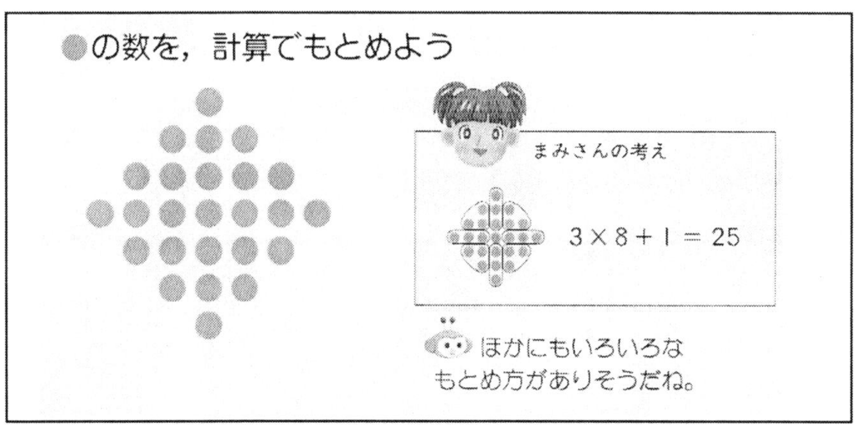

Source: Sugiyama, Iitaka, and Itoh (2002, p. 41).

Figure 12.2.   Children are asked to find different ways to calculate the total number of dots in the arrangement.

composite expression, that is, a mathematical expression with more than one operation. Such a discussion often starts with a problem like the following:

> Makoto had 1000 yen at first. He bought a 140 yen notebook and a 460 yen scissors. How much was the change? (Sugiyama & Hironaka, 2000c, p. 70)

This problem is accompanied by two different solutions, each involving two mathematical expressions. Then the book proposes, "Let's think about how to combine two math sentences into one and the order of calculations." To support students' thinking, textbooks often utilize "mathematical expressions with words." Starting with the following mathematical expression with words,

$$[\text{Total Payment}] - [\text{Price}] = [\text{Change}],$$

students then are expected to write the expression, $1000 - (140 + 460) = 400$, to describe the relationship among the quantities in this problem situation. Mathematical expressions with words are also used in various formulas in elementary schools. For example, the formula for the area of a parallelogram is given as *Area = Base × Height*, not *A = BH*, since students in intermediate grades have yet to study mathematical expressions with letters used as variables.

In the upper elementary grades (Grades 5 and 6), mathematical expressions with symbols such as □ , ○ , and △ are introduced and used more frequently. Students learn to express relations among quantities using these symbols in place of specific quantities. They then discuss questions like the following:

> In the above expression (□ × 4 = ○), what is ○ when □ is 10?

> What is □ when ○ is 28? (Sugiyama & Hironaka, 2000d, p. 58)

Although the 2000 COS moved mathematical expressions with letters from Grade 6 to Grade 7, the first year in lower secondary school, these examples show how central are the ideas related to *shiki* in the Japanese elementary mathematics curriculum.

KUFU

The word *kufu* is not a mathematics term. As a verb, this word is often translated as to devise, to contrive or to think out. Although each of these English words captures some aspects of the meaning of *kufu*, none seems

Source: Sugiyama and Hironaka (2000b, p. 35).

Figure 12.3. Children in Grade 2 are asked to kufu (figure out) all 7s facts.

to be a complete match. In general, this word means to devise something by modifying or applying what is already known in a novel manner. Thus, phrases like "to tinker with" or "to figure out" also capture some essences of the word. How is this word being used in the Japanese elementary mathematics curriculum?

Figure 12.3 shows a page from the introductory unit on multiplication in Grade 2. Students have already studied the multiplication facts with 2, 3, 4, 5, and 6 as the multiplicand, and they have identified some patterns among those facts. Here, they are asked to figure out (*kufu*) all 7s facts using what they have learned previously.

In Grade 4, after students learn about angles and how to measure angles using a protractor, students are asked to think about (*kufu*) how they can measure an angle that is greater than 180 degrees. Also in Grade 4, students investigate how they can simplify the division process when both the dividend and the divisor end with zero. After students learn that the quotient does not change when you eliminate the same number of zeros from both the dividend and the divisor, the textbook asks students to "think about an effective way of dividing 2700 by 400" (Sugiyama & Hironaka, 2000c, p. 37). Here again, students are asked to solve a problem similar to that with which they are familiar, but the method they developed previously cannot be applied in a straightforward manner.

These examples show that the notion of **kufu** is often invoked to encourage students to analyze what they have previously learned so that they can apply it in a novel problem—perhaps with some modifications. As the examples above suggest, *kufu* is often used to develop algorithms or procedures. Thus, algorithms and procedures are not just simply given to the students. Rather, students develop algorithms and procedures from simple cases, and then continuously refine them (*kufu*) as the problems become more complex.

POSSIBLE IMPLICATIONS FOR THE
U.S. ELEMENTARY MATHEMATICS CURRICULUM

What might be some implications of these examples for the U.S. elementary school mathematics curriculum? I would like to conclude this paper by raising three issues.

The first issue is the role of formal mathematical notation systems in (elementary) school mathematics. The emphasis of the Japanese curriculum on *shiki* suggests that the Japanese mathematics education community considers it important that students learn to express their thinking process as well as use mathematical notations to express relations among quantities. Most, if not all, U.S. mathematics educators also would agree that mathematical notations are important for students to learn. However, not many U.S. textbook series include discussions on how to write composite expressions or using words and mathematical symbols to express relationships among quantities. Mathematical expressions are used, but little attention is paid to helping students develop their ability to read and write mathematical expressions. There seems to be a tendency to shy away from discussing the use of formal mathematical notations in many of today's curricula. This is perhaps because we feel symbolic manipulations have been overemphasized in the traditional mathematics curricula of the past. The approach observed in the Japanese elementary school mathematics curriculum, however, seems to suggest a possible way to use mathematical symbols as both the objects and tools for learning.

The second issue I would like to raise is the role of particular algorithms and procedures in school mathematics. Treatment of traditional computational algorithms has been a point of contention in today's "math war" in the United States. Most, if not all, reform curricula emphasize student-invented algorithms. The question, "Should long division be taught?" symbolizes the current debate. In the Japanese curriculum, there is no question that the answer is "yes." From that perspective, the Japanese curriculum is a traditional curriculum.

Most U.S. curricula that emphasize student-invented algorithms encourage students to use their own reasoning ability to develop computational algorithms that make sense. Some do not even attempt to teach the conventional algorithms. There is an apparent belief that instruction of a particular, even non-traditional, algorithm and an emphasis on reasoning are mutually exclusive. However, the Japanese curriculum's emphasis on *kufu* while students are learning the traditional computational algorithms seems to suggest that it might be possible to develop a curriculum that teaches particular algorithms as a result of students' reasoning activities. In fact, as an example in Figure 12.2 suggests, it might even be possible to take advantage of children's diverse strategies to develop a particular understanding.

SANSU

The last issue I would like to raise deals with the other Japanese word I have not yet discussed: *sansu*. This word is different from the word, *sugaku*, which is used to label mathematics as an academic discipline, and is also used as the title of secondary school mathematics. According to Mr. Yoshikawa, the term *sansu* was created as the title of elementary school mathematics less than 100 years ago. Although I am not familiar with the origin of this word, it is interesting to note that a different word was invented instead of using the same word for mathematics or labeling it as "elementary school mathematics." Underlying this decision may be a belief that, although one important role of elementary school mathematics is to lay the foundation for the study of more advanced mathematics, *sansu* is an academic discipline in its own right.

The idea that *sansu* is something different from secondary school mathematics is also evident in the Japanese popular culture as well. When I was in Japan a few years ago, I noticed an advertisement in a train by a well known *juku* (cram school) which included a mathematics problem that is to be solved using only *sansu*. There are several Web sites where problems are posted weekly, again, to be solved using only *sansu* (for example, www.sansu.org). These sites seem to attract hundreds, perhaps thousands, of regular visitors who compete to be the first one to submit a correct solution to the weekly problem. Even though many Japanese mathematics educators are concerned about students disliking mathematics (*sugaku*), as indicated by TIMSS and other studies, Japanese people seem to have a much more favorable view of *sansu*.

These observations seem to suggest that it might be fruitful for mathematics educators to examine the nature and content of the elementary school mathematics curriculum. Through such an examination, we should try to articulate what should be the role of elementary school mathematics curriculum. If the goals of an elementary school mathematics curriculum include more than just preparing the students for more advanced mathematics, we need to articulate carefully and specifically what those goals are. A clearer articulation of goals of elementary school mathematics will also inform the way we prepare elementary school teachers. In the United States, the time seems to be ripe to have such a discussion. In Japan, such a discussion might offer an insight into how to deal with the issue of students disliking more advanced mathematics.

NOTE

1.   One potential challenge to doing similar work in a U.S. classroom is that there is no consensus on which factor in a multiplication sentence indicates the multiplicand. Thus, "3 × 8" may mean 3 sets of 8 or 8 sets of 3. In the Japanese elementary school curriculum, the first factor is considered the multiplicand; therefore, "3 × 8" always indicates 8 sets of 3.

REFERENCES

Hiebert, J., R. Gallimore, H. Garnier, K. Bogard Givvin, H. Hollingsworth, et al. (2003). *Highlights from the TIMSS 1999 video study of eighth-grade mathematics teaching*. Washington, DC: U.S. Government Printing Office.

Schmidt, W. H., Houang, R., & Cogan, L. (2002). A coherent curriculum: The case of mathematics. *American Educator, 26*(2), 10–26.

Schmidt, W. H., McKnight, C. C., & Raizen, S. A. (1996). *A splintered vision: An investigation of U.S. science and mathematics education*. Boston: Kluwer.

Sugiyama, Y., & Hironaka, H. (2000a). *New elementary mathematics 1* (in Japanese). Tokyo: Tokyo Shoseki.

Sugiyama, Y., & Hironaka, H. (2000b). *New elementary mathematics 2* (Vol. 2) (in Japanese). Tokyo: Tokyo Shoseki.

Sugiyama, Y., & Hironaka, H. (2000c). *New elementary mathematics 4* (Vol. 1) (in Japanese). Tokyo: Tokyo Shoseki.

Sugiyama, Y., & Hironaka, H. (2000d). *New elementary mathematics 4* (Vol. 2) (in Japanese). Tokyo: Tokyo Shoseki.

Sugiyama, Y., S. Iitaka, and S. Itoh. 2002). *New elementary mathematics 4* (Vol. 2) (in Japanese). Tokyo: Tokyo Shoseki.

Takahashi, A., Watanabe, T., & Yoshida, M. (2004). *Elementary school teaching guide for the Japanese course of study: Arithmetic (Grades 1-6)*. Madison, NJ: Global Education Resources.

Watanabe, T. (2007). Initial treatment of fractions in Japanese textbooks. *Focus on Learning Problems in Mathematics, 28*(1).

CHAPTER 13

A LOOK AT JAPANESE JUNIOR HIGH SCHOOL MATHEMATICS TEXTBOOKS

Blake E. Peterson
Brigham Young University

The purpose of this paper is to describe junior high school mathematics textbooks from Japan and highlight some interesting features in them as compared to textbooks in the United States. A complete analysis of all U.S. textbooks and Japanese textbooks was not done. However, based on my experiences observing lessons taught by student teachers and cooperating teachers in Japan, I identified aspects of the curriculum and the implementation of the curriculum that were different from my observations in the United States. Subsequently, a more detailed analysis of these aspects was conducted. First, the general organization and structure of the Japanese textbooks based on their tables of contents will be discussed. Next, the structure and flow of problems and examples in the first chapter of one of the Japanese seventh-grade mathematics textbooks will be investigated. Finally, some of the supplemental materials available to Japanese teachers will be examined.

Mathematics Curriculum in Pacific Rim Coutries—China, Japan, Korea, and Singapore:
Proceedings of a Conference, pp. 209–231

GENERAL ORGANIZATION AND STRUCTURE OF JAPANESE JUNIOR HIGH SCHOOL MATHEMATICS TEXTBOOKS

There are six publishers who publish mathematics textbooks used in Japanese junior high schools. Some of the general attributes of these textbooks are shown in Table 13.1. The attributes of some commonly used seventh-grade textbooks from the United States are shown in Table 13.2. The two reform oriented textbooks from the United States (Billstein & Williamson, 2005; Lappan, Fey, Fitzgerald, Friel, & Phillips, 1998) have fewer chapters and sections than the more traditional U.S. textbooks but are still significantly larger than those from Japan.

For a visual comparison of the Tokyo Shoseki seventh-grade book (Sugiyama, Fujiwara, & Morimoto, 2002a) and the Glencoe seventh-grade textbook (Collins et al., 1999), consider Figure 13.1. Two factors of the larger size textbooks in the U.S. are publishers desire to have a specified number of lessons and their attempts to craft one textbook that satisfies all teachers and all states' standards. Two contributing factors to the Japanese textbooks being so small is the expectation of parents that their students learn all of the content in the textbook and the careful scrutiny of

Table 13.1. Japanese Textbook Attributes

	Dai Nihon Tosho	*Gakkou Tosho*	*Keirinkan*	*Kyouiku Shuppan*	*Osaka Shoseki*	*Tokyo Shoseki*
Dimensions	6 × 8	6 × 8	6 × 8	6 × 8	7 × 10	6 × 8
Pages	187	199	183	211	167	195
Chapters	6	7	7	6	6	6
Section	17	15	30	16	15	15

Table 13.2. U.S. Textbook Attributes

	Glencoe (Mathematics Course 2)	*Glencoe (Pre-Algebra)*	*Holt Mathematics (Course 2)*	*McDougal Littell (Middle School Math Course 2)*	*McDougal Littell (Math Thematics Book 2)*	*Prentice Hall (Connected Math Project Grade 7)*
Dimensions	9 × 11	8.5 × 11	8.5 × 11	8.5 × 10.5	8.5 × 10.5	8 × 10
Pages	682	837	739	819	658	691
Chapters	13	13	12	13	8 modules	8 books
Section	98	101	102	90	40	48

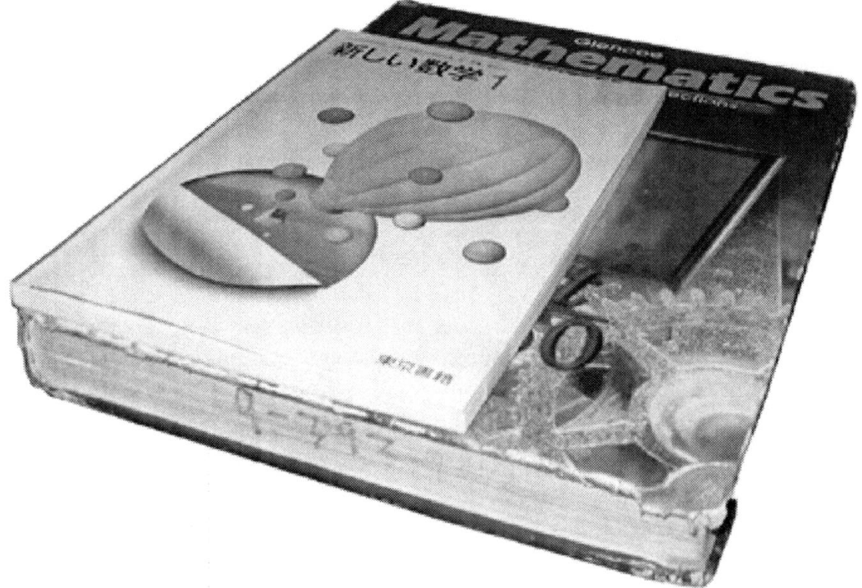

Figure 13.1.   A visual comparison of a Japanese and U.S. seventh-grade mathematics textbook.

the textbooks by the ministry of education (S. Yoshikawa, personal communication, November 11, 2005).

The tables of contents of the six textbooks are very similar in organization (Fukumori, Koseki, Morisugi, & Okamoto, 2002; Hiraoka & Yoshida, 2002; Ichimatsu, Okada, & Machida 2002; Masada, 2002; Sawada, Okamoto, & Sakai 2002; Sugiyama et al., 2002a). Four of the six textbooks have six chapters and two have seven chapters. Five of the six textbooks have between 15 and 17 total subsections and one textbook has 30 subsections. Many of the headings of the chapters and subsections are the same in the five textbooks with 15-17 subsections. The sixth textbook, published by Keirinkan (Fukumori et al., 2002), covers the same major topics in the same order as the rest of the textbooks but the subsections are broken up into smaller chunks. For example, the five textbooks with 15-17 subsections cover the topics of "positive and negative numbers" and "addition and subtraction" in sections 1 and 2 of chapter 1. The Keirinkan textbook, however, spreads this discussion across five subsections and two chapters. The subsections are titled "numbers smaller than zero," "expressing values with positive and negative numbers," "comparing positive and negative numbers," "addition and subtraction 1," and "addition and subtraction 2" (Fukumori et al., 2002).

The Tables of Contents

Although the Keirinkan textbook differs from the other five textbooks in the way that the subsections are structured, the main chapter headings are the same for all six textbooks. These headings closely parallel the seventh-grade content outlined in The Courses of Study in Japan shown below.

A.   Numbers and algebraic expressions

1.   To enable students to understand positive and negative numbers through activities in concrete situations and to perform the four basic fundamental operations.
2.   To enable students to develop the ability to express relationships and rules in mathematical expressions using letters. To enable students to grasp the meaning of expressions and to calculate mathematical expressions using letters.
3.   To enable students to understand and apply linear equations with one variable.

B. Geometrical figures

1.   To enable students to develop their ability to fully understand basic plane figures and to construct these figures.
2.   To enable students to examine geometrical figures through observation, manipulations, and experimentation and to understand further the concept of solid figures. To enable students to develop their ability to measure figures.

C. Mathematical Relations

1.   Through investigating the changes and correspondence of two quantities within the context of concrete phenomena, to enable students to develop their ability to discover, represent, and examine proportional and inversely proportional relationships. (Japan Research Team (JRT) for U.S.-Japan Comparative Research on Science and Mathematics Education, 2004, pp. 73–75)

Because of the similarity of all six textbooks, I chose to examine one textbook series carefully as a representative sample of the rest of the textbooks. The table of contents of this seventh-grade text, published by Tokyo Shoseki, is shown in Figure 13.2.

Chapter 1: Positive and negative numbers
  1. Positive and negative numbers
  2. Addition and subtraction
  3. Multiplication and division

Chapter 2: Letters and expressions
  1. Expressions using letters
  2. Simplifying expressions with letters

Chapter 3: Equations
  1. Equations
  2. First degree equation applications

Chapter 4: Proportions and inverse proportions
  1. Proportions
  2. Inverse Proportions
  3. Applications of proportions and inverse proportions

Chapter 5: Plane figures
  1. Symmetrical figures
  2. Construction of basic figures

Chapter 6: Solid figures
  1. Various 3-dimensional shapes
  2. Various ways of viewing 3-dimensional shapes
  3. Surface area and volume of 3-dimensional shapes

Source: Sugiyama et al. (2002a, pp. 4–5).

Figure 13.2.   Table of contents of the Tokyo Shoseki seventh-grade mathematics textbook.

The chapter headings of the tables of contents for Tokyo Shoseki's eighth and ninth-grade textbooks are shown in Figures 13.3 and 13.4 respectively. The complete tables of contents for both books are shown in Appendix A. These tables of contents in conjunction with the seventh-grade table of contents give a sense of the flow of mathematical ideas during all three years of junior high school.

Some Observations About the Table of Contents

Traditionally the curriculum in the U.S. has compartmentalized mathematics content into pre-algebra, algebra, geometry, algebra II and pre-calculus. One obvious observation is the integrated nature of the Japanese curriculum which is similar to many of the National Science Foundation-funded reform curricula. Not only are two of the chapters in each of the seventh, eighth, and ninth-grade textbooks devoted to geometry but it is

Chapter 1: Simplifying expressions

Chapter 2: Simultaneous equations

Chapter 3: First-degree functions

Chapter 4: Parallelism and congruence

Chapter 5: Characteristics of figures

Chapter 6: Probability

Source: Sugiyama et al. (2002b, pp. 4–5)

Figure 13.3.   Chapter headings of the table of contents of the Tokyo Shoseki eighth-grade mathematics textbook.

Chapter 1: Square Roots

Chapter 2: Polynomials

Chapter 3: Second-degree equations

Chapter 4: Similar figures

Chapter 5: Pythagorean Theorem

Chapter 6: Functions

Chapter 7: Various problems (mathematics applications and ways of thinking about mathematics)

Source: Sugiyama et al. (2002c, pp. 4–5)

Figure 13.4.   Chapter headings of the table of contents of the Tokyo Shoseki ninth-grade mathematics textbook.

also quite common for teachers to use geometric situations as a context to build linear and quadratic equations.

A second observation about the organization of the curriculum is the careful use and introduction of language. For example, the term "linear" in conjunction with equations or functions is not introduced until high school or college. Instead linear equations are referred to as first-degree equations in the seventh grade. In the eighth grade, first-degree functions are introduced but "linear" is still not used. Similarly, the term "quadratic" is not used to describe equations of the second degree. Instead they are called second-degree equations. In The Courses of Study in Japan outlined above, item number 3 under numbers and algebraic expressions says "linear equations with one variable" (JRT, 2004, p. 73).

The literal translation, however, is "one variable, first-degree equations," which is the same as the uses in the tables of contents in all six textbooks.

Another example of the use of vocabulary deals with integers. The set of integers is introduced at the end of elementary school and is the focus of the first chapter in all six seventh-grade mathematics textbooks. However, as the operations on positive and negative numbers are taught, the term integers is only briefly mentioned. Subsequent references use the descriptive language of positive and negative numbers (*seifu no suu*). Although *seifu no suu* is a broader set than integers because it includes decimal representations like –0.8 and rational numbers like $-\frac{2}{3}$, it is a simpler vocabulary than integers or rational numbers.

The third example of simplified vocabulary is the use of the words letter or character for variable. In Japanese the word *moji* means letter or character. When variables are introduced in the Japanese textbooks, they are not introduced with a new vocabulary word but are introduced with the common word *moji*. This term is used throughout the seventh-grade textbook. In fact there does not appear to be a separate term for variable used in the junior high school curriculum.

It is understood that there are four meanings of "*moji*" in Japan, as

1. constant but unknown number (numbers in general)
2. constant number, such as "*e*" or "π"
3. unknown letter (used in equations)
4. variable (H. Ninomiya, personal communication, November 29, 2005).

In each of these cases of vocabulary use, a simpler more common word is used to introduce a concept. This common terminology is used for more than just an introduction but is used throughout most of the junior high school curriculum.

NATURE AND SEQUENCING OF EXAMPLES AND PROBLEMS

To gain a more detailed understanding of the flow of ideas in a specific section of the textbook, I worked with a native Japanese speaker to translate the first 20 pages of the seventh-grade textbook published by Tokyo Shoseki (Sugiyama et al., 2002a). The main components of the sections in this textbook are questions (*hatsumon*), examples, solutions, problems and exposition. I describe and provide examples of each of these components to give a sense of the organization and development of ideas in this textbook.

Hatsumon

Almost every section begins with a deep, thought provoking question sometimes called a *hatsumon* in Japanese. The Japanese term, *hatsumon*, which means *"asking a key question that provokes students' thinking"* (Shimizu, 1999, p. 109) has no direct translation into English. There are a few sections that do not start with a *hatsumon* and some other sections have a second or third *hatsumon* in the middle of the section. Some examples of *hatsumons* are provided below.

On page 6 the first statement in the book is a *hatsumon* which says "Let's investigate how to express temperature." This statement is followed with a question to help students investigate the *hatsumon*: "On the map find the city with the highest temperature and the lowest temperature. What is the difference between these temperatures?" As we will discuss later, it is not the intent of this question for the students to do subtraction but rather to look at the temperatures that are displayed on the map of the world and think about the meanings of the different numbers. The original *hatsumon* is further refined with the question "What does Moscow's –3° represent?"

The *hatsumon* on page 15 is used in the introduction of addition and subtraction of positive and negative numbers. Up to this point, the text has only discussed how to represent positive and negative numbers and has nothing written about operations with positive and negative numbers. This *hatsumon* is shown in Figure 13.5.

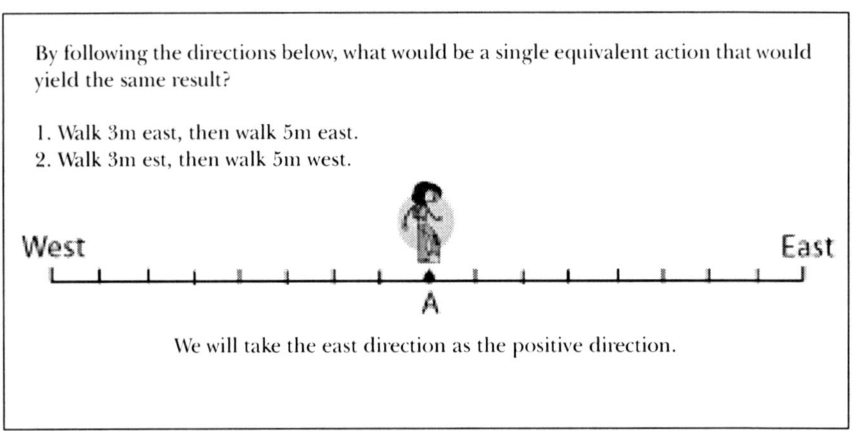

Source: Sugiyama et al. (2002a).

Figure 13.5.   The *hatsumon* found on page 15.

> Calculate the expressions in A and B and compare the results.
>
> **A.** $\{(+3) + (-9)\} + (+7)$
> **B.** $(+3) + \{(-9) + (+7)\}$

Source: Sugiyama et al. (2002a).

Figure 13.6.   The *hatsumon* found on page 19.

A third *hatsumon* is located in the middle of section 2 and follows a discussion of addition of two positive or negative numbers. It is used to motivate the extension of the associative property of addition from just positive numbers to positive *and* negative numbers (see Figure 13.6).

Examples

Many examples in U.S. mathematics textbooks are simply examples of how to solve problems. A problem is stated and a solution to the problem is presented. In the first few pages of the Tokyo Shoseki seventh-grade mathematics textbook, the examples are different in nature than what we typically see in the United States. Because the mathematical topic in these pages is about positive and negative numbers, the "examples" are descriptions of cases where these numbers are used. The following three examples are of that type.

Example 1 (page 9): If 500 yen income would be represented as $+500$ yen, then -300 yen would represent spending 300 yen.

Example 2 (page 9): If we set sea level as 0 meters in altitude and anything above sea level is positive, then the height of Mt. Fuji is represented by $+3776$m.

Example 4 (page 10): On the map below the expected highest temperature for various cities are shown. The number in the parentheses represents the change in temperature from the previous day. For example, the (-2) by Sendai indicates the expected highest temperature for the day is $2°$ lower than the highest temperature of the previous day. (Sugiyama et al., 2002a)

The next example (see Figure 13.7) is more computationally oriented but is focused on how to think about the computation as opposed to how to do the computation. (Note: Because the example numbering starts

Example 3: Addition of numbers with the opposite sign.

1. (-5) + (+5)
    = 0

2. (+9) + (-4)
    = + (9 − 4)
    = +5

3. (+4) + (-10)
    = -(10 − 4)
    = -6

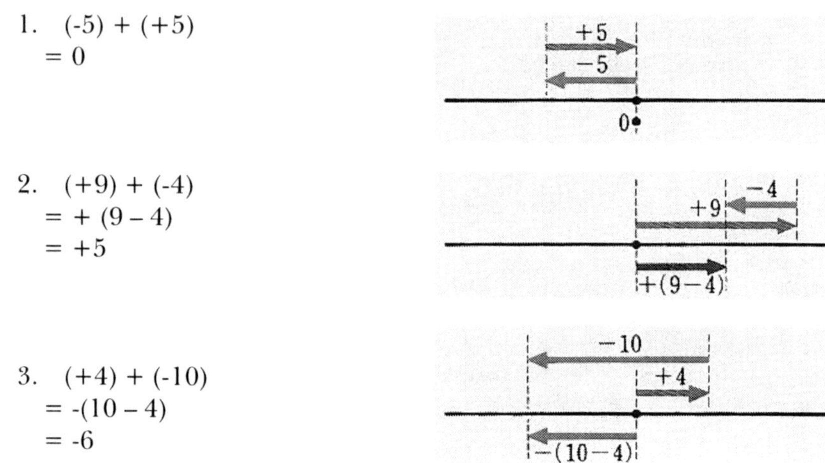

Source: Sugiyama et al. (2002a).

Figure 13.7.   Example 3 is found on page 17.

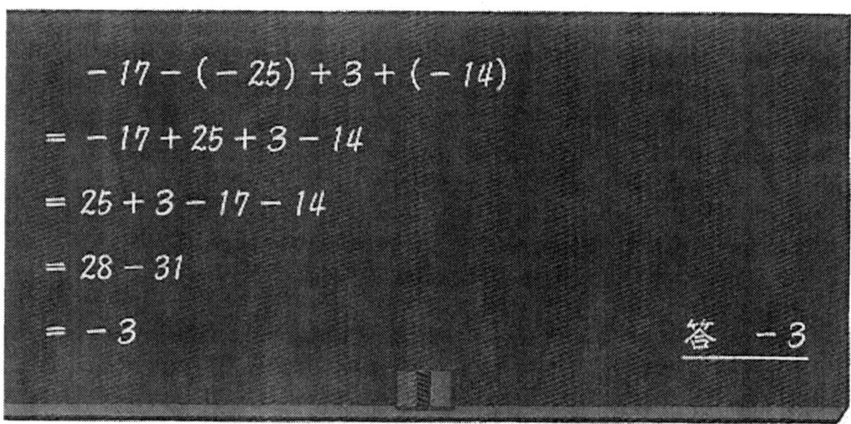

Source: Sugiyama, Fujiwara, and Morimoto (2002a).

Figure 13.8.   Example 1 is found on page 24.

over in each new section, Example 4 on page 10 precedes Example 3 on page 17.)

The last example (see Figure 13.8) is more typical of what would be seen in a textbook in the Unites States. The example consists of a problem and a solution. The solution to the problem is shown on a chalkboard to indicate that this is an exemplary way of writing the solution.

Problems

The largest portion of the text consists of a combination of examples and problems. With the occasional *hatsumon* to initiate a new idea, problems and examples are interwoven together to develop the target concept. There is some exposition also mixed in but the majority of the ideas are developed through examples and problems. The most interesting aspect of this use of problems to develop concepts is that the solutions to the problems are usually not presented. Thus the students would, of necessity, have to gain some confidence in the validity of their solutions.

In Example 2 on page 9, discussed previously, the students are given an example of how positive and negative numbers are used to represent altitudes above and below sea level. This example is followed by problem 2 (see Figure 13.9) in which the students are asked to represent the altitudes of a couple of locations in the country: no solutions are provided.

Problem 4 on page 10 also closely follows the concept in the example that precedes it. Example 4 on page 10 describes a map of the country with predicted temperatures and changes in temperatures. Problem 4 (see Figure 13.10) follows example 4 and asks for the meaning of the positive and negative numbers on the map.

Problem 5 on page 10 (see Figure 13.11) highlights a concept similar to that in problem 4 but is not related to any particular example.

Problem 2: Following the example, use (+) and (-) signs to represent the following altitudes.
1. Nobeyama train station's altitude is 1345.67m.
2. The seikan tunnel is 240m below sea level.

Source: Sugiyama et al. (2002a).

Figure 13.9.   Problem 2 is found on page 9.

Problem 4: What do Sapporo's (0) and Fukuoka's (+2) represent?.

Source: Sugiyama et al. (2002a).

Figure 13.10.   Problem 4 is found on page 10.

Problem 5: If we set the altitude of Mt. Yarigadake, which is 2903m, as the standard and any heights greater are represented with (+), and any height lower represented with (-), then how would you represent the following?
    a. Shiroumadake 2932m.
    b. Shakushidake 2812m

Source: Sugiyama et al. (2002a).

Figure 13.11.   Problem 5 is found on page 10.

Problem 2: Do the following computations.
    1. (+4) + (–3)   **2.** (+7) + (–9)
    3. (+6) + (–6)   **4.** (–12) + (+6)

Source: Sugiyama et al. (2002a).

Figure 13.12.   This Problem 2 is found on page 17.

The problems presented thus far are focused on the conceptual meanings of positive and negative numbers and what they might represent. Some problems, however, are just computation. Example 3 on page 17 (see Figure 13.7) introduces the addition of numbers with opposite signs and Problem 2 (see Figure 13.12), which follows right after, asks students to do computations of addition. (Note: Similar to examples, the numbering for problems starts over in each new section.)

In this case as in all other cases that I investigated, there were no solutions presented to the problems. In some ways, problems like those above which are interwoven among the examples function like "homework" problems one might find at the end of a section in a U.S. textbook. Even though there are not solutions to these problems in the book, there are different books available in bookstores that look similar to the textbooks but contain a bit more explanation and include the solutions to the problems. These "solution books" are not provided by the schools but are available for the students to purchase if they desire.

"Homework"

At the end of each section there are a few exercises for students to do. There are also additional chapter problems at the end of each chapter.

Section 1 contains four exercises at the end of the section. The following two exercises from Sugiyama et al. (2002a, p. 14) are examples of those at the end of section 1.

1. Fill in the blank. If 8 minutes from now is written as +8 minutes, 5 minutes ago would be written as ______.

4. Within each of the following, determine which is larger and use an inequality sign to express that.

1. +3, –4   **2.** –13, –8   **3.** 0, –6, +3

Summary

Almost every section in the Tokyo Shoseki seventh-grade mathematics textbook begins with a *hatsumon* to introduce the new concept of the section. This is followed by a combination of examples and problems woven together to develop the concept. There is also some exposition in each section but the largest portion of the text is a mixture of the examples and problems.

In section 1 of chapter 1, which addresses the concept of positive and negative numbers, there are two *hatsumon*, five examples, and 12 problems with four exercises at the end of the section. In the entire seventh-grade book there are 62 *hatsumon*, 76 examples, and 192 problems. The total number of exercises at the end of sections and chapters is 109. Because most of these review exercises have more than one part, it is worthwhile to consider that the 109 exercises have an average of 2.8 parts.

By comparison, the Glencoe seventh-grade textbook (Collins et. al, 1999) uses exposition and examples to illustrate concepts. This textbook also includes about four labs in each chapter. These labs are more exploratory in nature but are presented in such a way that they could easily be ignored. Each of these labs has about nine problems in it. At the conclusion of each section there is a set of exercises consisting of practice problems, application problems and a mixed review. At the end of each chapter there are some review problems and test problems. Altogether, the Glencoe book contains 299 examples, 3,128 exercises, 51 labs containing 486 problems and 928 review problems.

SUPPLEMENTAL MATERIALS

In every language and culture, what is written in a textbook may be quite different from the actual curriculum enacted in the classroom. For this reason, I also investigated some of the supplemental materials available to students and teachers. These materials are student workbooks, ancillary books and teacher guidebooks.

Student Workbooks

Student workbooks are commonly available for student to use in conjunction with their regular textbooks. These workbooks and the textbooks that they supplement are not necessarily published by the same publisher. For example, there are at least three different workbooks published by different publishers that correspond to the seventh-grade mathematics textbook published by Tokyo Shoseki (Sugiyama et al., 2002a).

These workbooks have a large number of problems and exercises and are closely correlated to the content of the textbook. Figure 13.13 includes an example page of the workbook published by Shingakusha (Iwasaki, 2002). The top of each page in this workbook includes the textbook page numbers to which the workbook problems correspond. There is also an additional explanation of the concepts introduced in those textbook pages at the top of the workbook page. The rest of the page consists of groups of problems. Each set of problems has a heading of the example or problem that it corresponds to in the textbook (see Figure 13.13).

Ancillary Materials

Accompanying the Tokyo Shoseki textbooks is a brochure for various supplemental materials. The titles of two of the books referenced in this brochure were *The Method of Opening the Window to Mathematics: Hints for Junior High Lessons* (Nozaki, 1995) and *Lessons That Elicit Thinking: Arithmetic, Mathematics* (Handa, 1995). An examination of the tables of contents of these two books (shown in Figure 13.14 and 13.15) indicates that the purpose is not to provide specific activities or tasks for specific lessons in the text or in the middle school curriculum but to discuss approaches to mathematics lessons in general.

Each chapter in *The Method of Opening the Window to Mathematics: Hints for Junior High Lessons* focuses on a general pedagogical issue but the sections within the chapters refer to specific mathematical concepts. The chapter titles for this book are shown in Figure 13.14. A complete table of contents is found in Appendix B.

The chapters in *Lessons That Elicit Thinking* emphasize the development of lessons in broad mathematical categories. The chapter titles for this book are shown in Figure 13.15. A complete table of contents is found in Appendix B.

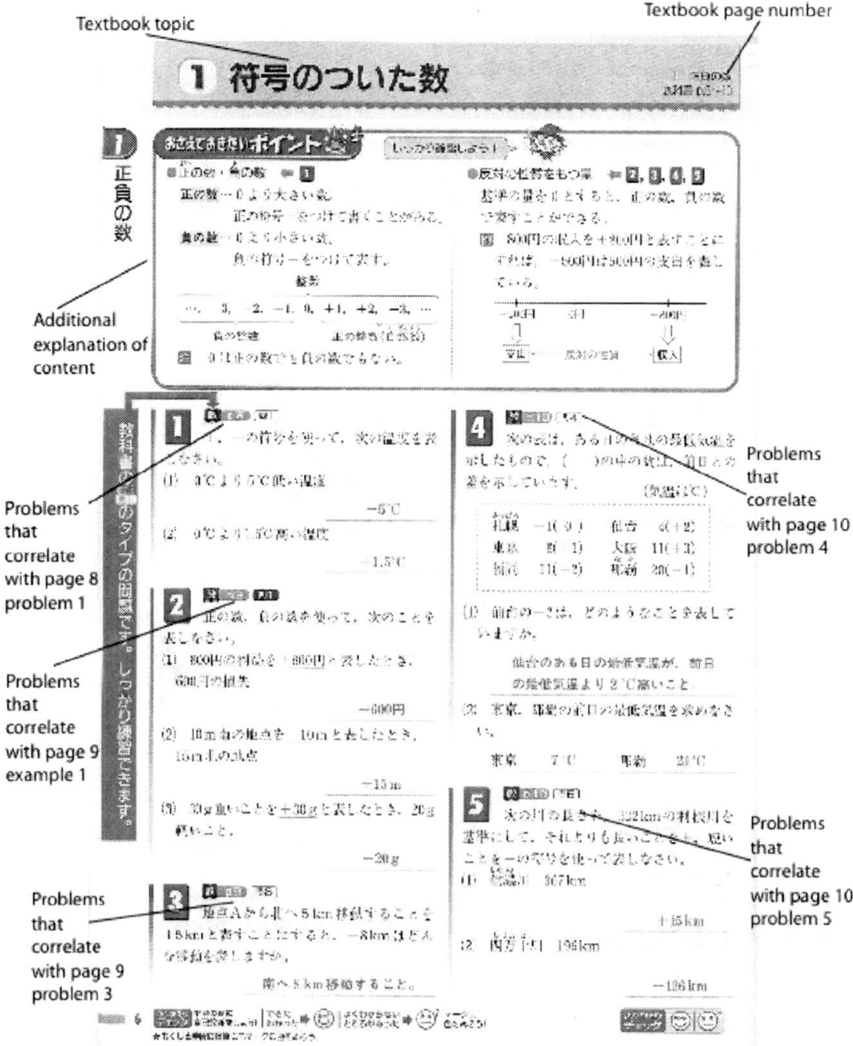

Source: Iwasaki (2002).

Figure 13.13. Page 6 of the workbook to accompany the seventh-grade mathematics textbook published by Tokyo Shoseki.

Teacher's Guidebook

The last supplemental material that I examined was the teacher's guidebook (Tokyo Shoseki, 2002) for the seventh-grade mathematics textbook published by Tokyo Shoseki (Sugiyama et al., 2002a). Unlike the

---

Chapter 1: Your mathematics, my mathematics

Chapter 2: Hand made mathematics

Chapter 3: Living mathematics

Chapter 4: Making explanations that are easily understood by students

Chapter 5: Designing lessons

Source: Nozaki (1995).

Figure 13.14.   Table of contents for *The Method of Opening the Window to Mathematics: Hints for Junior High Lessons*.

Chapter 1-Designing a lesson which elicits thinking

Chapter 2-A lesson which elicits thinking about numbers and equations

Chapter 3-A lesson which elicits thinking about the use of diagrams

Chapter 4-A lesson which elicits thinking about the use of functions

Chapter 5-A lesson which elicits thinking about probability and statistics

---

Source: Handa (1995).

Figure 13.15. Table of contents for *Lessons That Elicit Thinking: Arithmetic, Mathematics*.

ancillary books described above, this guidebook is very closely correlated with the student textbooks. Similar to a teacher's edition of a textbook found in the United States, the teacher's guidebook has a small copy of the pages of the student textbook with the supporting materials written around it. There is a different guidebook for each textbook grade level. The initial pages of the seventh-grade guidebook were translated. These pages were selected because they corresponded to the translated pages of the student textbook.

The first item listed under each section or subsection is a list of aims or goals. The section beginning on p. 8 with a title "Numbers with Signs" has the following goals:

- Through an example of the temperature, be interested in negative numbers, and think about the meaning of negative numbers and 0.

- To try to find an example from things around you where positive and negative numbers are used.

- To understand the meaning of positive and negative numbers.

- To understand that 0 is the standard of quantity.
- To try to represent the changes or quantities that have opposite values by using positive and negative numbers.
- To be able to represent the changes or quantities that have opposite values by using positive and negative numbers.
- To represent whether a number is bigger or smaller using positive and negative numbers after you set a standard (Tokyo Shoseki, 2002, p. 20).

The subsection beginning on p. 11 focuses on the size and ordering of numbers. The goals for this subsection are:

- To try to think about a number line that has negative numbers.
- To be able to represent positive and negative numbers on a number line.
- To be able to come up with a value after you understand the meaning of an absolute value.
- To be able to understand size of positive and negative numbers, and to represent it by using inequality signs. (Tokyo Shoseki, 2002, p. 23)

After listing the goals of the section or subsection, the guidebook goes on to discuss the various components. Key conceptual points may be discussed or additional detail supporting a certain example or problem might be presented. For example, the opening paragraph, supporting the first pages of the textbook, discusses a critical underlying concept for students to grasp negative numbers. It says:

When we use negative numbers, we need to review the meaning of 0. From elementary school, the students know that zero means nothing. Some elementary school teachers teach about negative numbers, but that is insufficient. If students see 0 as nothing, it is difficult for them to understand that there are numbers less than 0. Students must review the meaning of 0 and numbers less than 0. They need to be able to understand that there are numbers less than 0.

If we pick any number as a standard, then it is natural to see that there are numbers that are greater than and numbers that are less than the standard. If we set 0 as our standard, the numbers that are greater than, we call positive numbers, and the numbers that are less than we call negative. In order to use negative numbers it is necessary to instruct students about the meaning of 0 which is the standard upon which positive and negative numbers based.

> As we have students think about he meaning of 0, it is important to clarify how 0 is represented as the standard. (Tokyo Shoseki, 2002, p. 18)

This theme of setting 0 or some other number as the standard is further emphasized in the discussion that supports some of the subsequent examples and problems. In the explanation of example 2 on p. 9 it says, "It is an example that a location, which is above the surface of the ocean (sea level), is expressed by + since the standard of 0 is the surface of the ocean. In this case, a location which is under the surface of the ocean is expressed by −" (Tokyo Shoseki, 2002, p. 20). The guidebook goes on to further elaborate that altitude is based on the average sea level. In Japan, the standard of the sea level is established to be at a certain spot in the bay near Tokyo.

In problem 2 on p. 9 (see Figure 13.9), the students are asked to express the altitude of Nobeyama station and the tunnel between the main island and Hokkaido. The guidebook explains why these locations might be of interest to the students. Nobeyama is significant because the highest point in the entire Japanese railway system is within a few kilometers of this station. The location and details of the tunnel are also explained in the guidebook.

CONCLUSION

In conclusion, this examination of Japanese seventh-grade mathematics textbooks provides ideas for U.S. textbook authors to consider. The magnitude of U.S. textbooks is overwhelming whether it be in the size of the book or in the sheer number of practice exercises. More concise textbooks with fewer practice exercises should be considered. One way that this could be done would be to develop concepts through questions and problems as seen in the Japanese textbooks rather than extensive exposition.

A second idea that could be considered from the Japanese textbooks is the use of language. Using first-degree equations instead of linear equations is an example of using simpler, very descriptive language to describe a mathematical object instead of introducing a more abstract term. Are there other examples of mathematical terms in English that could be replaced with less abstract, more descriptive words?

Finally, the supplemental materials provided to teachers can be more than a set of quickly implemented activities but can develop the pedagogical theory of teaching mathematics and the rationale behind certain practices. Such materials could enrich not only the content of a specific lesson but impact the daily practice of teachers. This pedagogical theory could be in the form of developing the underlying mathematical concepts in ways that are consistent with the development of the problems in the textbook. While some of the National Science Foundation-funded curricula has some of this supporting pedagogical theory, teachers would benefit if all textbooks had more of this type of supplemental material.

The nature of the Japanese supplemental materials in conjunction with the organization of the textbooks, highlight the fact that simply importing textbooks from other countries may have little impact on the mathematics instruction in the United States. Providing teachers with the supporting supplemental materials that clarify the theory and rationale of the textbooks is also an important part of impacting what goes on in the classroom.

APPENDIX A: TABLE OF CONTENTS FOR EIGHTH- AND NINTH-GRADE MATHEMATICS TEXTBOOKS PUBLISHED BY TOKYO SHOUSEKI

New Mathematics 2

Chapter 1: Simplifying expressions
    1. Simplifying expressions
    2. Applications of variable expressions
Chapter 2: Simultaneous equations
    1. Simultaneous equations
    2. Applications of simultaneous equations
Chapter 3: First degree functions
    1. First degree functions
    2. First degree functions and equations
Chapter 4: Parallelism and congruence
    1. Parallel lines and angles
    2. Congruent figures
Chapter 5: Characteristics of figures
    1. Triangles
    2. Parallelograms
    3. Triangles and circles
Chapter 6: Probability
    1. Probability

Additional Material
- Review [make sure of content]
- Mathematics independent research problems
    (i) How many times does toilet paper turn
    (ii) Celsius and Fahrenheit

    (iii) Geometry and Euclid
    (iv) Arrangement of figures to create traditional tessellations
    (v) Let's read Braille (Sugiyama et al. 2002b).

New Mathematics 3

Chapter 1: Square roots
    1. Square roots
    2. Simplifying radical expressions
Chapter 2: Polynomials
    1. Calculating (expanding) polynomials
    2. Factoring polynomials
Chapter 3: Second-degree equations
    1. Methods of solving second-degree equations
    2. Applications of second-degree equations
Chapter 4: Similar figures
    1. Similar figures
    2. Parallel lines and ratios
Chapter 5: Pythagorean Theorem
    1. Pythagorean Theorem
    2. Pythagorean Theorem applications
Chapter 6: Functions
    1. Functions
Chapter 7: Various problems (mathematics applications and ways of thinking about mathematics)

Additional Materials
- Review
- Mathematics independent research
    Problems listed on page 173
    (i) How far does a lighthouse's light reach?
    (ii) Using the carpenter's tool called sashigane {square}
    (iii) Measuring height {using methods of old} {math text from edou era}
    (iv) Golden ratio
    (v) Let's think with origami
    (vi) Let's use functions to think about global warming (Sugiyama et al., 2002c).

APPENDIX B:
TABLE OF CONTENTS FOR ANCILLARY BOOKS

The Method of Opening the Window of Mathematics: Hints for Junior High Lessons

Chapter 1: Your mathematics, my mathematics
  Section 1: The secret of the multiplication table (9 by 9 multiplication table)
  Section 2: Spirograph
  Section 3: Making problems that require the building equations
  Section 4: Drawing on the coordinate plane
  Section 5: Let's draw similar characters
  Section 6: Report using cartoons
  Section 7: Making a picture book of mathematics
Chapter 2: Hand made mathematics
  Section 1: Lining up the factors of 72
  Section 2: Cards (integers) adding to zero version of "Concentration"
  Section 3: Holding hands, two straight lines
  Section 4: Guessing the number using the process of 2s
  Section 5: Dividing oil (Pouring oil from bucket to bucket to get a specified amount)
  Section 6: Represent roots using continued fractions.
Chapter 3: Living mathematics
  Section 1: Using a balance scale to talk about functions
  Section 2: Diagram
  Section 3: Environment and functions (looking at tables and graphs).
  Section 4: How can we learn about 3000 years ago?
  Section 5: How much is the shipping cost?
  Section 6: The centroid of triangles and squares.
  Section 7: Pascal's probability tree
Chapter 4: Making explanations that are easily understood by students
  Section 1: Proof of Diagram
  Section 2: From making a square to Pythagorean Theorem
  Section 3: Explain Pythagorean Theorem by using cut out puzzle
  Section 4: Delta polyhedron and perfect (even) polyhedron
  Section 5: A mysterious relationship in the perfect (even) polyhedron
Chapter 5: Designing Lessons
  Section 1: A mysterious world of star-shaped polygons
  Section 2: Think about "Hatome ageshi"

Section 3: A map of proposition
Section 4: Solution of formula by tiles
Section 5: Analysis of variation of quantity (Nozaki 1995).

Lessons That Elicit Thinking: Arithmetic, Mathematics

Chapter 1-Design a lesson which elicits thinking
    Section 1-A lecture which elicits thinking
    Section 2-Structure of a lecture which elicits thinking
Chapter 2-A lesson which elicits thinking about numbers and equations
    Section 1-A goal for teaching numbers and equations
    Section 2-Expansion of the world of numbers and development stage of children
        -Example of actual performance of teaching numbers
    Section 3-Guidances to progress the work of equation
Chapter 3-A lesson which elicits thinking about the use of diagrams
    Section 1-A goal for the teacher guidance diagrams
    Section 2-Teacher guidance about diagrams in elementary school
    Section 3-Consideration and application of characteristics of diagrams
Chapter 4-A lesson which elicits thinking about the use of functions
    Section 1-Nature of functions and a lesson which makes you think
    Section 2-Development of consciousness of functions
    Section 3-Teacher guidance to support the eye for functions
    Section 4-Teacher guidance for linear functions
Chapter 5-A lesson which elicits thinking about probability and statistics
    Section 1-An intention of probability and statistics lectures
    Section 2-Teacher guidance to grow the way to look at probability and statistics (Handa, 1995).

REFERENCES

Billstein, R., & Williamson, J. (2005). *Middle grades math thematics*. Evanston, IL: McDougal Littell.

Bennett, J. M., Burger, E. B., Chard, D. J., Jackson, A. L., Kennedy, P. A., Renfro, F. L., et al. (2007). *Holt mathematics: Course 2*. Austin, TX: Holt, Rinehart, and Winston.

Collins, W., L. Dritsas, P. Frey, A. C. Howard, K. McClain, D. Molina, et al. (1999). *Mathematics applications and connections: Course 2*. New York: Glencoe/McGraw-Hill.

Fukumori N., Koseki, N., Morisugi, K., & Okamoto, K. (2002). *Suugaku ichinen* [Mathematics: First year]. Tokyo: Keirinkan.

Handa, S. (1995). *Kangae saseru jugyo: Sansuu suugaku* [Lessons that elicit thinking: Arithmetic, mathematics]. Tokyo: Tokyo Shoseki.

Hiraoka, T., & Yoshida, M. (2002). *Chuugakkou suugaku 1* [Junior high school mathematics 1]. Tokyo: Dai Nippon Tosho.

Ichimatsu, M., Okada, I., Machida, S. (2002). *Chuugakkou suugaku 1* [Junior high school mathematics 1]. Tokyo: Gakkou Tosho.

Iwasaki, M. (2002). *Shin suugaku waku 1: Tokyo Shoseki* [New mathematics workbook 1: Tokyo Shoseki]. Tokyo: Shingakusha.

Japan Research Team (JRT) for U.S.-Japan Comparative Research on Science and Mathematics Education. (2004). *The courses of study in Japan*. Tokyo: Japan Ministry of Education.

Lappan, G., Fey, J. T., Fitzgerald, W. M., Friel, S. N., & Phillips, E. D. (1998). *Connected mathematics: Grade 7*. New York: Prentice Hall.

Larson, R., Boswell, L., Kanold, T., & Stiff, L. 2005. *Middle school math: Course 2*. Evanston, IL: McDougal Littell.

Masada, M. (2002). *Chuugaku suugaku 1* [Junior high mathematics 1]. Osaka: Osaka Shoseki.

Nozaki M. (1995). *Suugaku no mado no akekata: Chugaku no jugyo e no hinto* [The method of opening the window of mathematics: Hints for junior high lessons]. Tokyo: Tokyo Shoseki.

Sawada, T., Okamoto, M., & Sakai, Y. (2002). *Chuugaku suugaku 1* [Junior high mathematics 1]. Tokyo: Kyouiku Shuppan.

Shimizu, Y. (1999). Aspects of mathematics teacher education in Japan: Focusing on teachers' roles. *Journal of Mathematics Teacher Education, 2*(1), 107–116.

Sugiyama, Y., Fujiwara, M., & Morimoto, M. (2002a). *Atarashii Suugaku 1* [New mathematics 1]. Tokyo: Tokyo Shoseki.

Sugiyama, Y., Fujiwara, M., & Morimoto, M. (2002b). *Atarashii suugaku 2* [New mathematics 2]. Tokyo: Tokyo Shoseki.

Sugiyama, Y., Fujiwara, M., & Morimoto, M. (2002c). *Atarashii suugaku 3* [New mathematics 3]. Tokyo: Tokyo Shoseki.

Tokyo Shoseki. (2002). *Atarashii suugaku 1: Kyoushiou shidousho shidouhen* [New mathematics 1: Teacher's guidebook]. Tokyo: Author.

CHAPTER 14

EXPLORING KOREAN PRIMARY MATHEMATICS

Janice Grow-Maienza
Truman State University

Susan Beal
University of Illinois at Chicago

Of great interest to American researchers and mathematics educators about Asian primary mathematics textbooks is how those texts conceptualize fundamental mathematics for children (Grow-Maienza, 2002: Grow-Maienza & Beal, 2003, 2005; Mayer, Sims, & Tajika, 1995; Li, 2005; Watanabe, 2001). Our study of Korean mathematics began with an observation study in 32 Korean first- and fifth-grade mathematics classes in 1995-96 (Grow-Maienza, Hahn, & Joo, 2001), the period in which the 6th national curriculum was in place in Korea (Ministry of Education, Korea 1993), and Korean students ranked second on the Trends in International Mathematics and Science Study (TIMSS) (U.S. Department of Education, 1996).

In the process of analyzing data from that study, the investigators found that teachers observed had adhered to the national curriculum content. Examination of a sample translation of the textbook revealed the

Mathematics Curriculum in Pacific Rim Coutries—China, Japan, Korea, and Singapore: Proceedings of a Conference, pp. 233–253

same results as had the classroom observations. Lessons were characterized by systematic development of mathematical procedures tied to their conceptual underpinnings. With a grant from the National Science Foundation, the series (12 student texts and 12 teacher manuals for Grades 1 through 6) was translated at Truman State University, and the English translation of the 6th national curriculum is called here *Korean Mathematics* (2001). Because of the close similarity to lessons observed in the observation study, and because of Korea's high ranking on the 1995 TIMSS, we assume we are working with what was in 1995 the intended, taught, and attained curriculum in Korea.

The texts are rich, indeed, and so we are pleased to have this opportunity to share them with you. The materials are concise and systematic, presented in a block structure. Each concept is developed concretely, and moves to the abstract, presenting sequentially more complex problems. The teacher manual states at least two objectives for every lesson—the first and most important objective is to understand a given principle, the second to apply that understanding to a procedure or operation. Every lesson starts with a meaningful story problem. Manipulatives and number lines are used widely, even into the fifth-grade lessons. The most salient characteristic of *Korean Mathematics* is that every procedure and operation is tied to its underlying mathematical principle.

To demonstrate how *Korean Mathematics* builds number sense for children, in this paper we show examples of how operations and procedures are tied to two principles—first, the inverse relationships of operations, and second, the base ten place value system that is taught from the very beginning in primary mathematics. Next, we demonstrate examples of how multiple models are used to reinforce conceptualization, and how concepts build one upon the other. Finally, we will compare the in-depth development of one concept—the angle—across grade levels in *Korean Mathematics* with the development of that concept in two American curricula, a traditional and a reform curriculum.

INVERSE RELATIONSHIP OF OPERATIONS

Addition and subtraction are introduced in first grade with a chapter on merging and separating numbers. Children learn that 3 separated into 2 and 1 is the inverse of 2 and 1 merged into 3. Thus, children learn from the beginning that addition and subtraction have an inverse relationship.

The same principle is learned in third grade when children learn various ways to divide, all of them explicitly showing division to be the inverse relation of multiplication.

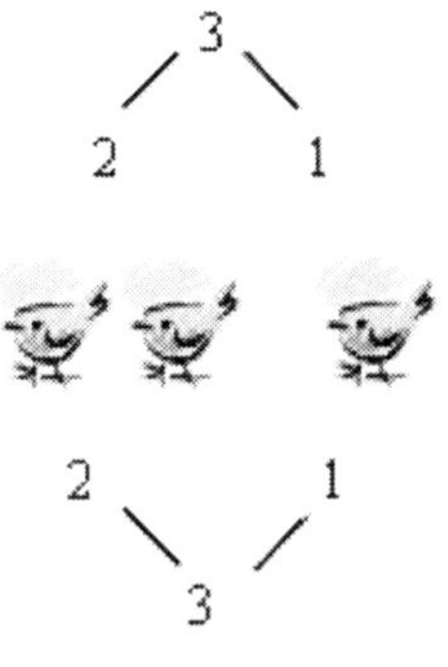

Figure 14.1.   Subtraction, the inverse relation of addition.

· In the second division lesson, a method for getting the quotient from the multiplication tables is introduced and explained thus:

$$\overline{}$$

?   ?

32 ? 8 = □   8 × □ = 32

?   ?

$$\underline{}$$

· To find the quotient in 32 ? 8,
8 × 4 = 32 is needed from the multiplication table.

Figure 14.2.   Division, the inverse relation of multiplication.

BASE 10 PLACE VALUE SYSTEM

A second major difference in *Korean Mathematics* from other textbook series we have examined is that from the beginning study of number, addition and subtraction are taught focused on the decimal place value

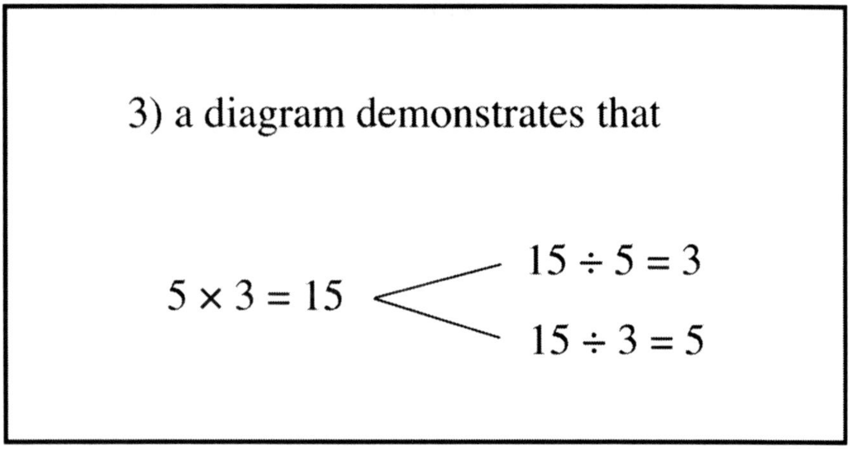

3) a diagram demonstrates that

$$5 \times 3 = 15 \quad < \quad \begin{array}{l} 15 \div 5 = 3 \\ 15 \div 3 = 5 \end{array}$$

Figure 14.3.   Division the inverse relation of multiplication.

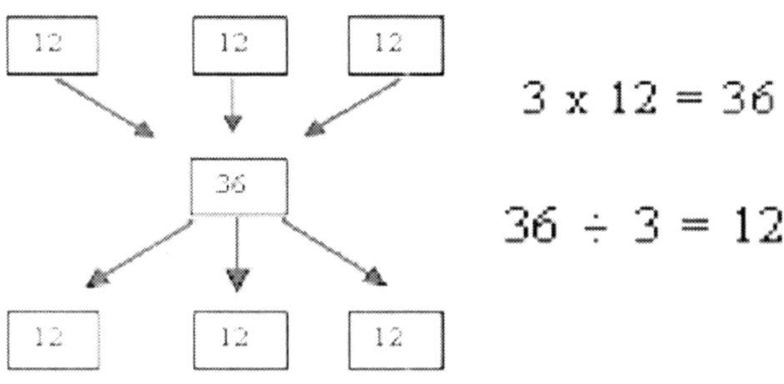

Source: Korean Mathematics (book 3-1-6, p. 80).

Figure 14.4.   Division, the inverse relation of multiplication.

system. Students using the *Arithmetic* (Ministry of Education, Korea 1993) series (which is a literal translation of the Korean textbook we have entitled *Korean Mathematics*) are not expected to learn facts beyond sums of 10. Children are first taught numbers that sum to 5, then those that sum to 10. Then, if adding 5 + 7, for instance, children base their responses on knowledge of sums of 10, separating the larger addend (7) into 5 + 2, adding 5 to 5 to make 10, and then adding 2, the remaining portion of 7, to 10 to make 12. Counting up and counting down to 10 that children learn in primary school, reported by Fuson and Kwon (1992a, 1992b), illustrates how children's learning of numbers is tied to the base 10 place value system from the beginning of their study.

Two ways of subtracting a one-digit number from a number up to 19 with renaming (borrowing) are also taught in *Korean Mathematics*: the subtraction-addition method, and the subtraction-subtraction method. Both methods involve separating the larger digit into a ten and ones. Children learn to subtract units from two-place numbers counting down to 10. In the subtraction-addition method the minuend is decomposed into a group of 10 and ones (in Figure 14.6, 3 ones) and the subtrahend (in Fig-

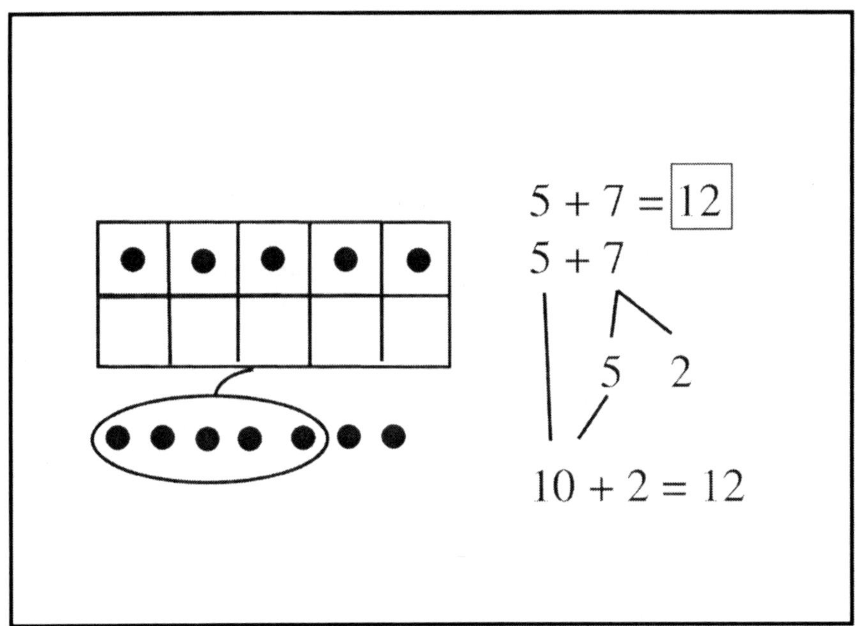

Source: Korean Mathematics (book 1-2-4, p. 46).

Figure 14.5.   Building base 10 number sense with addition.

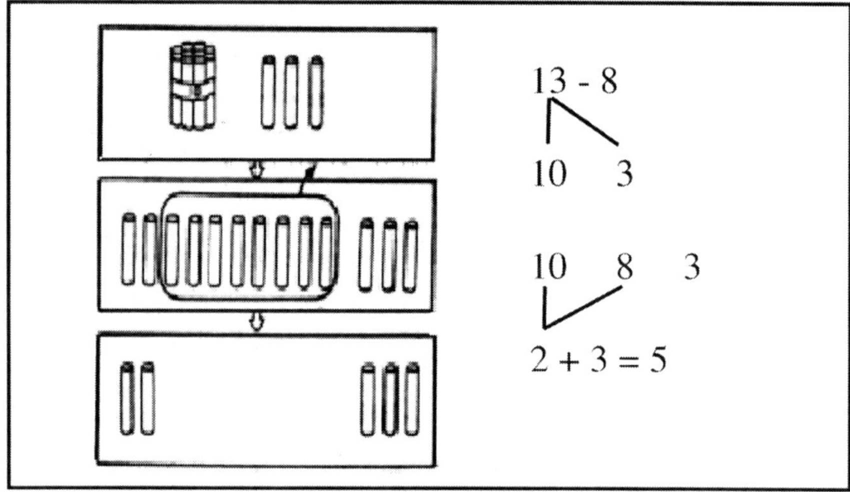

Source: Korean Mathematics, Teacher Manual (book 1-2-4, p. 89).

Figure 14.6.   Building base 10 number sense with (subtraction/addition) subtraction.

ure 14.6, 8) is subtracted from the group of 10. The difference (in Figure 14.6, 2) is then added to the 3 ones decomposed from the original two-digit number.

The subtraction-subtraction method of subtracting with renaming (borrowing) is taught, like the first method, through the manipulation of objects such as sticks, tooth picks, or *Ba-Duk* stones. The subtrahend is decomposed into numbers one of which, when subtracted from the minuend, results in 10. Then the remainder of the subtrahend is subtracted. Thus, in the calculation of 13 – 7, 7 is decomposed into 3 and 4. The 3 is subtracted from 13, resulting in 10, and then 4 is subtracted from 10.

According to the teacher manual, both subtraction methods must be learned to fluency, and then the student may choose. Note that these concepts, as shown here from a student text, are all presented with word problems, manipulatives, diagrams, and two different expressions of the algorithm, horizontal and vertical.

OPERATIONS MODELED ON THE NUMBER LINE

Multiple strategies are used in *Korean Mathematics* to teach these underlying principles. One model is the ubiquitous number line. Tad Watanabe demonstrated at the National Council of Teachers of Mathematics

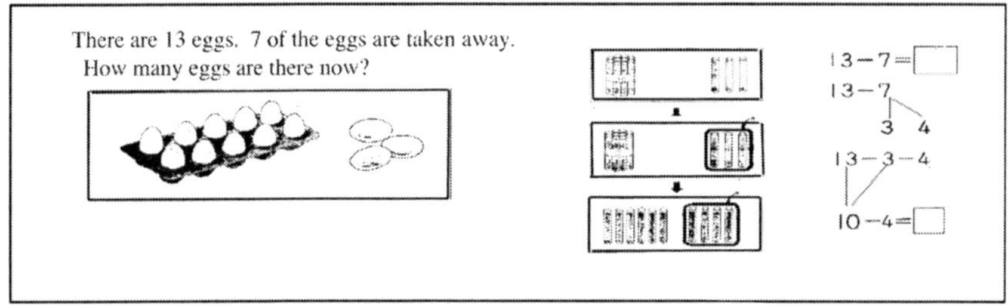

Source: Korean Mathematics (book 1-2-4, p. 51).

Figure 14.7.   Building base 10 number sense with (subtraction/subtraction) subtraction.

Research Presession in 2005 that use of the double number line in Japanese texts seems to be quite intentional on the part of the publisher (Grow-Maienza & Beal, 2005). So can the consistent use of the number line as a model seen throughout the Korean curriculum be seen as very intentional on the part of the curriculum developers in Korea. The number line is used as the first or second model in almost every operational concept. Figures 14.8 through 14.16 show operations modeled on the number line in the first- through the sixth-grade texts in *Korean Mathematics.*

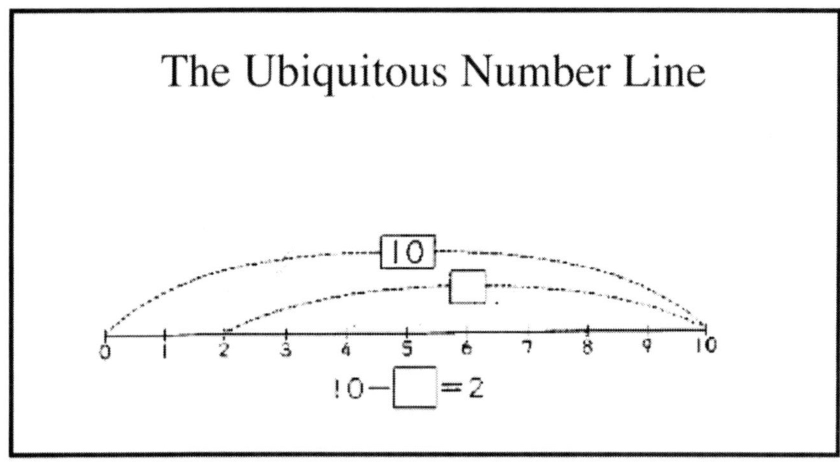

Source: Korean Mathematics (book 1-2-1, p. 9).

Figure 14.8.   Basic subtraction model.

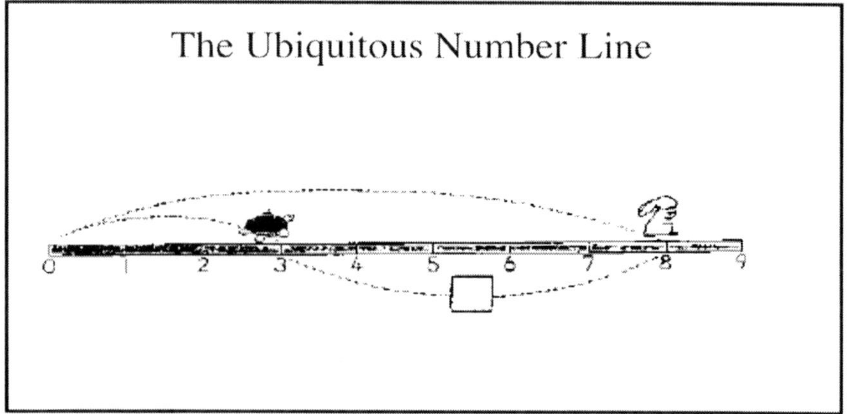

Source: *Korean Mathematics* (book 1-5-1, p. 65).

Figure 14.9.   Comparison subtraction model.

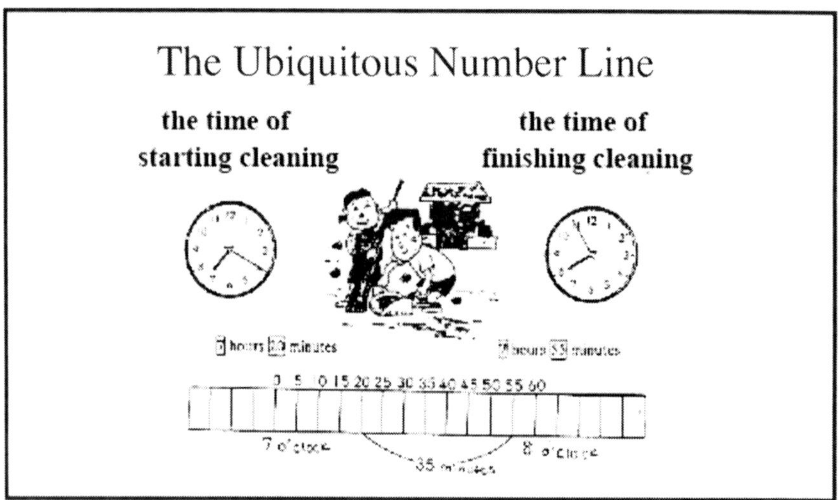

Source: *Korean Mathematics* (book 2-1-6, p. 62).

Figure 14.10.   Time modeled on number line.

Figure 14.8 shows basic "take away" subtraction, and Figure 14.9 shows the comparison model of subtraction as modeled on the number line. Figure 14.10 models "time" on the number line. Figure 14.11 models comparison of fractions with decimals. Figures 14.12 and 14.13 model the addition of subtraction of decimals. Figure 14.14 models the division of a

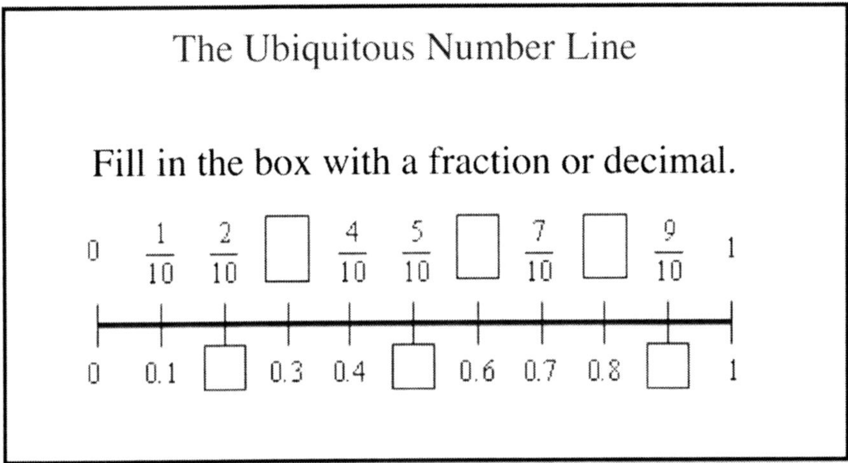

Source: *Korean Mathematics* (book 3-2-7, p. 95).

Figure 14.11.   Comparison of fractions with decimals.

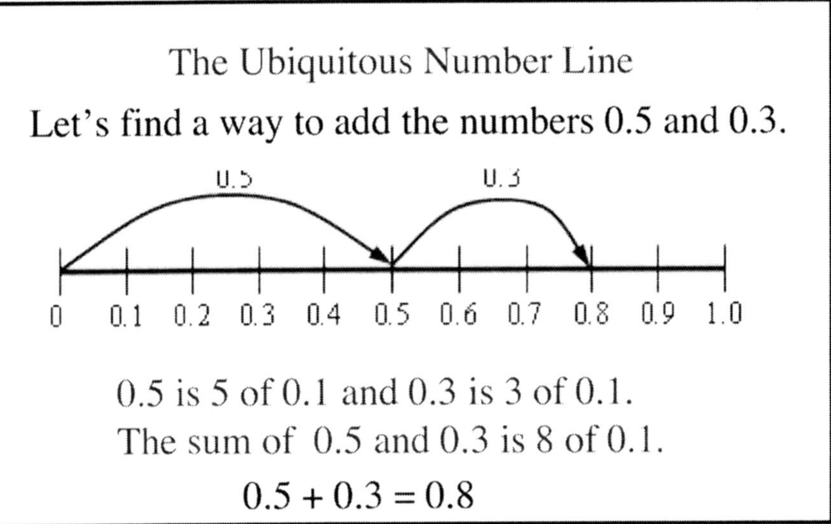

Source: *Korean Mathematics* (book 3-2-7, p. 96).

Figure 14.12.   Addition of decimals.

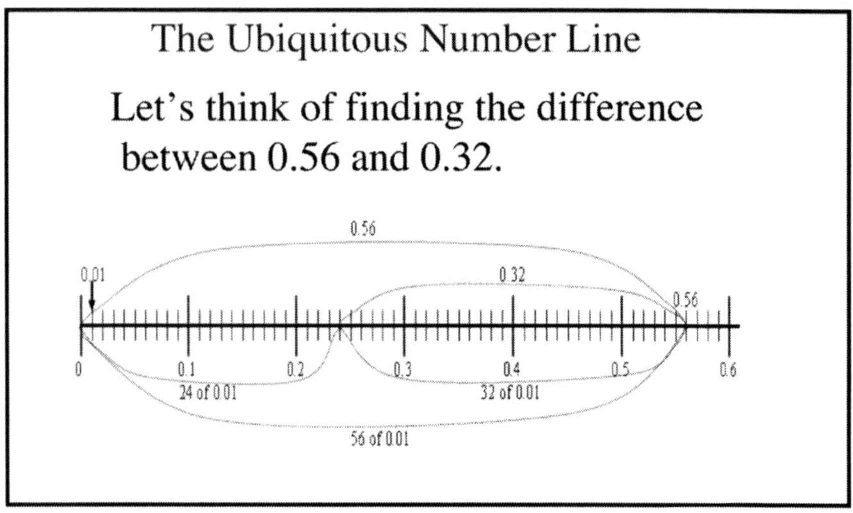

Source: Korean Mathematics (book 4-2-6, p. 78).

Figure 14.13.   Subtraction of decimals.

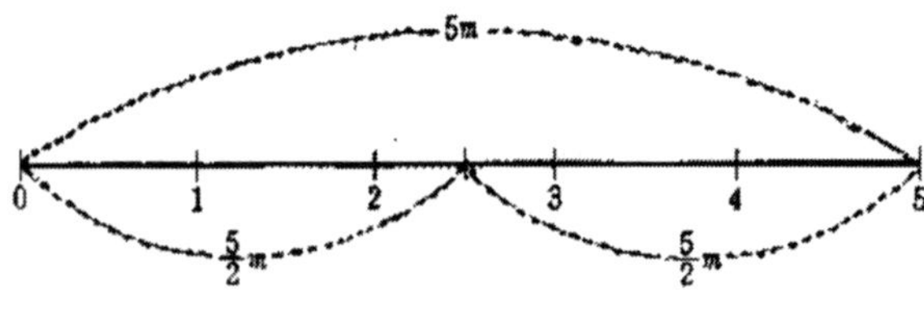

Source: Korean Mathematics (book 5-2-2, p. 26).

Figure 14.14.   Division of whole into fractions.

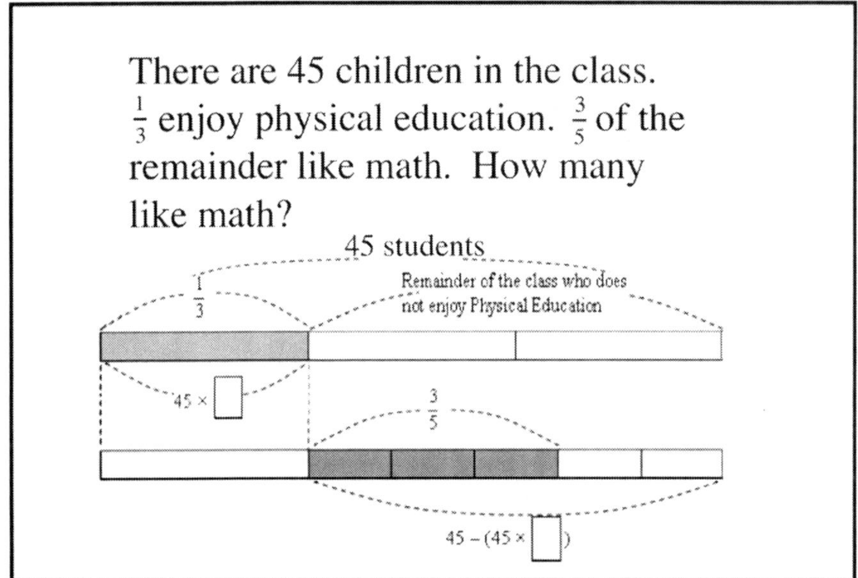

Source: *Korean Mathematics* (book 6-2-9, p. 133).

Figure 14.15.   Three step, mixed operations.

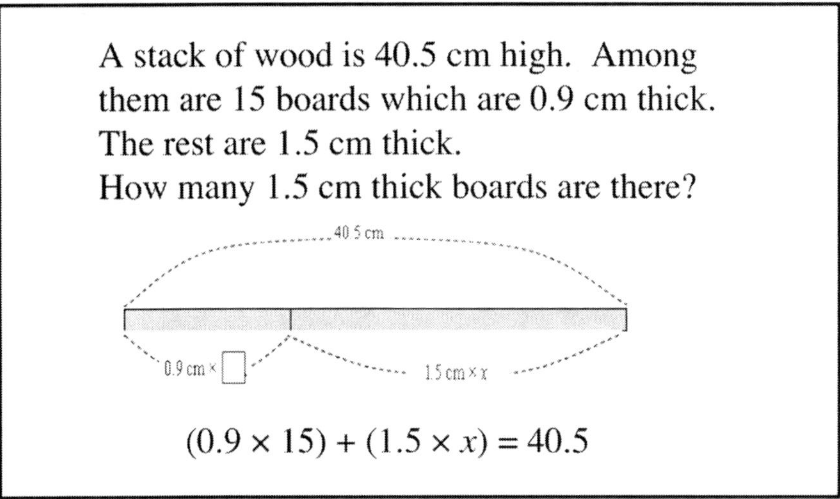

Source: *Korean Mathematics* (book 6-2-9, p. 135).

Figure 14.16.   Identification of *x*.

whole into fractions. And Figures 14.15 and 14.16 demonstrate three-step mixed operations and the identification of *x* on number lines.

BUILDING SEQUENTIALLY FROM CONCRETE TO ABSTRACT

In a classroom observation study in Pusan, Grow-Maienza and colleagues (2001) found that the observed lessons consisted of sequences of instruction/practice/evaluation, each instructional sequence addressing a more complex problem. Such patterns are being reported in other Asian classroom studies. Wang (2002) analyzed four Chinese beginning middle school teachers' lessons on triangles and their curriculum materials and found "they could use increasingly sophisticated mathematics problems to engage their students in integrating the current and previously learned concepts and providing justification for their problem solutions. Teachers moved from stages of instruction to guided practice, then to independent practice" (Wang & Lin, 2005, p. 6).

Grow-Maienza and Beal (2003) found similar patterns in Korean textbooks. Figures 14.17–14.19 from the fifth-grade textbook chapter on coordinate systems and graphs demonstrate how children are led from concrete to abstract thinking with increasingly complex problems. First introduced to a pictorial of an authentic coordinate field, the children are sequentially taught the construction of a table of coordinates and its

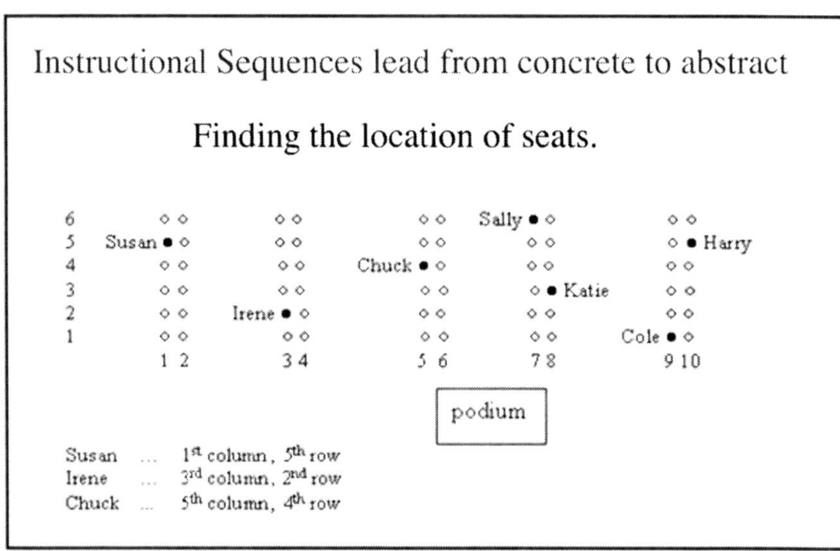

Figure 14.17.   Coordinate pairs on an authentic plane.

Relationship in table transposed to an equation

The rate of the amount of water flowing from a faucet is 3 L per 1 minute. When the amount of water is y and the time is x, find the comparative chart between x and y. Let's make a chart by filling in x and y.

x (min)	1	2	3	4	5	6	7	8
y (l)	3	6	9	12	15			

Which is called a relationship between x and y?
y is 3 times x in the above chart. Therefore, $y =$ ☐

Source: *Korean Mathematics* (book 5-2-7, p. 104).

Figure 14.18.   Construction of equation to show relationship of x and y.

graph on a coordinate field and then immediately taught to translate to an equation. The problems become more and more complex.

The sequence of problems below was observed by Grow-Maienza, and colleagues in their classroom observation study (2001). The tables and graphs of the equations demonstrate how teachers build on concepts in their instructional sequences. Figures 14.20–14.22 illustrate the graphs drawn in the class observed.

$y = x$ (graphed in Figure 14.19)

$y = \frac{1}{2}x$ (graphed in Figure 14.20)

$y = \frac{1}{2}x - 1$ (graphed in Figure 14.21)

$y = \frac{1}{2}x - 1$, (graphed in Figure 14.22; $y = \frac{1}{2}x$ graphed on dotted line)

Students constructed the table, graph and equation for $y = x$, and then were asked to construct the coordinate table, graph, and equation for $y = x$.

State the location of the following coordinate points as an ordered pair.

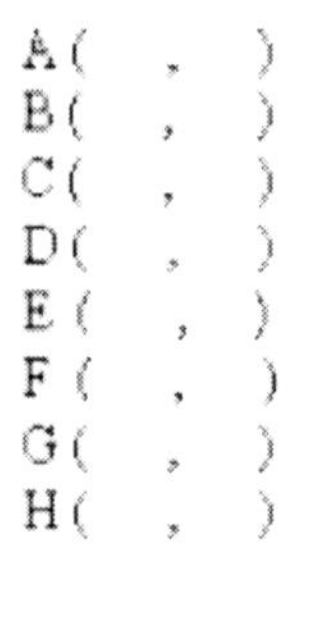

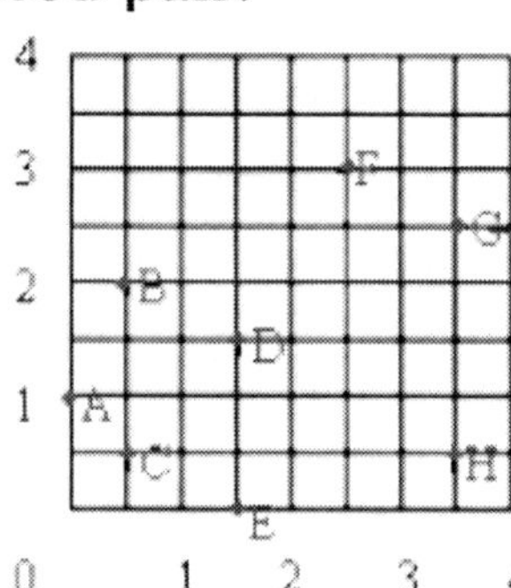

If x changes to 0, 1, 2, 3, 4, 5, the relation $y = 1 \times x$ has the following the chart and graph

$$y = 1 \times x$$

x	y	coordinates
0	0	(0,0)
1	1	(1,1)
2	2	(2,2)
3	3	(3,3)
4	4	(4,4)
5	5	(5,5)

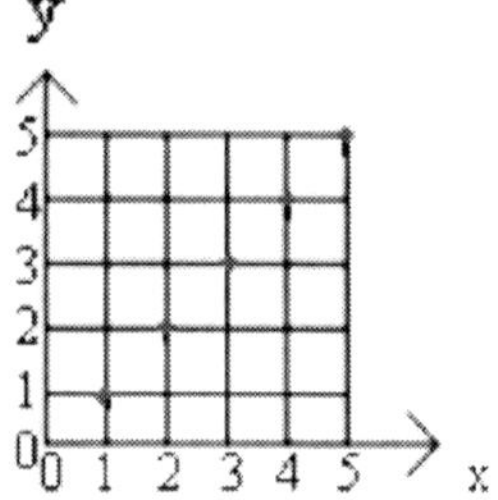

Source: *Korean Mathematics* (book 5-2-7, pp. 101, 106).

Figure 14.19.   Integration of coordinate pair table, plotted graph, and equation.

Students were then presented with the graph in Figure 21, and asked to identify the equation. After much problem exploration at their desks, students were asked to think of all the operations—addition, subtraction, multiplication, division. The teacher then drew the line representing $(y = x)$ on the plane to give students a clue to $(y = x - 1)$ as seen in Figure 22.

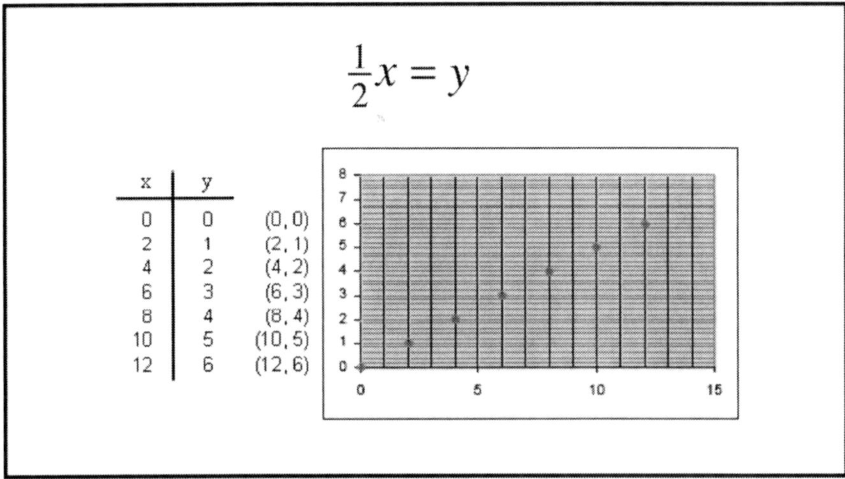

Figure 14.20.   Building on earlier concepts.

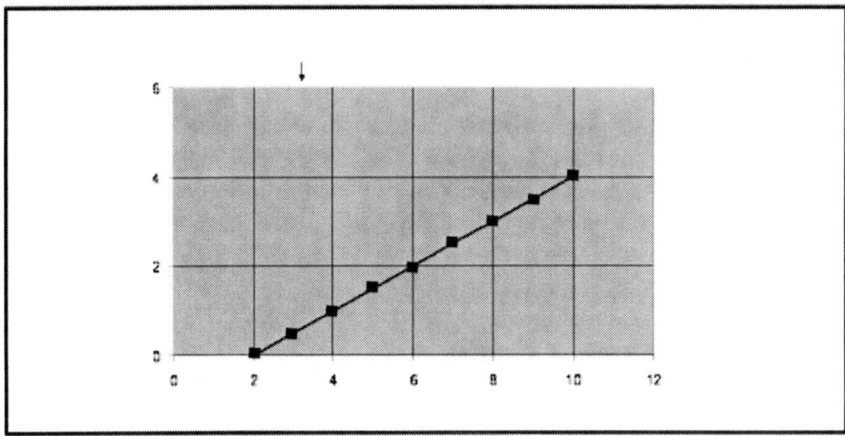

Figure 14.21.   Building on earlier concepts.

DEVELOPMENT OF THE CONCEPT OF ANGLE IN THREE ELEMENTARY SCHOOL CURRICULA

The concept of an angle is an integral part of learning about space and direction (orientation and motion). Angles also help to define the shape of objects. Furthermore, an angle is an important mathematical tool for describing and analyzing physical space. The angle also plays an impor-

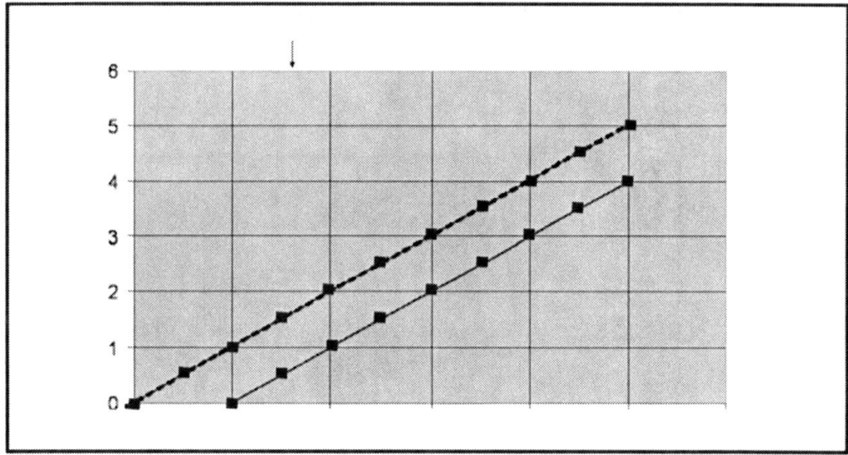

Figure 14.22.   Building on earlier concepts.

tant role in the development of geometry and trigonometry later in the students' education.

We wanted to compare the intended curriculum of *Korean Mathematics* (KM) with "typical" curricula from the United States. It was decided to look at *Everyday Mathematics* (EM) (2002) as an example of an NSF-funded curriculum. *Math Steps* (MS), published by Houghton Mifflin (2000), was chosen as an example of a traditional (for lack of a better adjective) curriculum. Both curricula state that they meet the NCTM Standards in Geometry.

Concept Development, Angles as Shape

All curricula gave students, at an early age, the opportunity to identify angles in two-dimensional figures and three-dimensional geometric solids, and angles in the students' environment. For the most part this was determined from the teacher manuals rather than a particular lesson, although in KM students did have activities to find shapes with three and four corners and draw shapes with three and four sides. MS also had students identify the number of corners and sides in one of the lessons. In EM, three or four students made shapes with a rope. However, both of these activities were not explicitly addressing the concept of angles. When describing the rope activity, the emphasis was on constant perimeter. And when describing the body activity, the emphasis was on the number of students needed on each side of the quadrilateral.

All three curricula examined the characteristics of solids, counting faces, edges and corner points, and students in KM were ask to construct a die and nets for rectangular solids. In KM there were also activities for tracing faces of solids, thereby connecting three-dimensional solids to two-dimensions.

Angles as Action, Turning Around a Center Point

All three curricula used the idea of rotations. EM and MS introduced quarter turns, half turns, three-quarter turns and full turns. It should be noted that MS did not explicitly connect the idea introduced in one lesson to the study of angles. The turns were explicitly connected to a circle measuring 360°. KM and EM used objects (sticks and straws, fastened), emphasizing the idea of an action and connecting it to the "size" of the angles made. MS did show a picture of what could be sticks that were fastened to represent an angle.

Both KM and EM took advantage of the hands on the clock to "discuss" the change of magnitude of angles. EM also turned the hands both clockwise and counterclockwise.

Angles as the Union of Two Rays

All three curricula defined angles as the union of two rays.

It is difficult, of course, from examining the textbooks and teacher manuals to determine if students are able to tie these three ways of approaching angles together to enrich their concept of angle. Only in KM, which has a block structure rather than a spiral structure, did students experience these ideas in close proximity.

Terminology

Both KM and EM used the terminology of quadrangles when emphasizing the study of angles, as compared to quadrilaterals. KM did so in the second half of the first grade when students were asked to construct triangles and quadrangles with sticks, and continued to use it when appropriate. EM introduced the terminology to the students in Grade 4 when talking about angles, triangles and quadrangles.

Naming Angles (Including Length of Rays)

All the curricula instructed students how to name angles. In the teacher manual for KM teachers were instructed to vary the length of the rays drawn so that students would be less likely to think that angles with longer rays were bigger. This caution was not found in the teacher manuals from EM or MS.

Establishment of the Right Angle

The right angle was typically introduced as a square corner. Angles were compared to the right angle. In this manner, acute and obtuse angles were introduced. MS also likened a right angle to 9 o'clock, and used the clock to talk about angles less than or greater than a right angle.

The straight angle was introduced in both KM and EM, and EM introduced the reflex angle. MS briefly mentioned the straight angle in the sixth grade when summing the angles of a triangle.

KM introduced the right triangle (and the regular triangle, i.e., equilateral triangle), and students drew right triangles with a right triangle ruler. MS classified triangles by their angles, thereby establishing right triangles, acute triangles and obtuse triangles.

Comparison of Angles Prior to Measuring Angles

KM was the only curriculum that had students compare angles prior to measuring the angles. First they had students overlap angles to determine which was larger. Then they introduced *arbitrary unit angles* with which to measure other angles.

Establishment of a Need for a Universal Unit

Using an arbitrary unit angle helped KM establish a need for a universal unit—1°. EM had students make arbitrary angles to measure various angles to also establish the need for a universal unit. The degree unit was presented to students in MS.

Measuring of Angles in Degrees Using a Protractor

All three curricula "taught" students how to use a protractor, how to read a measure of an angle, and how to draw an angle of a specified measure. EM was the only curriculum that had students construct their protractors, first a full-circle protractor in the third grade, and then a half-circle protractor in the fourth grade

Measure and Draw Angles with a Protractor

All curricula had students measure and draw angles with a protractor. KM did so at Grade 4, EM at Grades 5 and 6, and MS at Grades 6 and 7.

Circle Measures

An activity in EM used the 360° of a circle to find the measures of the angles in the various pattern blocks. Also in EM, a connection was re-established between right angles and the measure of the circle as 360 degrees. Percentages were applied to angles in the circle. In both EM and KM central angles were introduced to students in the sixth grade.

Calculating With Measures of Angles

KM had students add and subtract measures of angles as well as look at the sum of two angles forming a straight line in the fourth grade. No other curricula ask the students to do calculations with measures of angles. However, in the sixth grade, EM asked students to find measures of angles that summed to a straight angle without a protractor. MS only talked about supplementary angles in the sixth grade.

Finding measurements of angles were extended to explore concepts of perpendicular lines, intersecting lines, parallel lines cut by a transversal, supplementary and complementary angles, and relationships among the sum of angles in triangles, quadrilaterals, and polygons. In fact, measuring angles in a polygon, and summing the angles was a popular activity in all curricula. MS was the only curriculum that had students "find" a formula for the sum of the angles for polygons by triangulating the polygon.

Constructions Using Straight Edge and Compass

In the fifth grade KM had students study congruent figures. Students were to construct congruent triangles given another triangle, construct a triangle given measures of two angles and a side, as well as given measures of three sides. Students were also asked to find conditions when it is not possible to construct triangles. In the seventh grade of MS, students construct angles congruent to a given angle, as well as construct angle bisectors (and segment bisectors).

Extension to Other Geometrical Concepts

In KM, one of the ways symmetry was introduced to the students used measures of angles and lengths of sides.

The development of the concept of an angle was explored in three ways: angles as shape, angle as an action, and angles as the union of two rays. All three curricula used these three "definitions" of an angle in their lessons, but none explicitly helped students connect these definitions. KM came closest to doing so by virtue of the structure of their curriculum, that is to say, by the block structure, rather than the structure of their lessons (Grow-Maienza & Beal, 2003).

The concept development of angle in KM preceded the measuring of angles. This was not true of the other curricula in that concept development and measurement were presented, at times, in tandem. All three curricula taught students how to measure angles with a protractor, but EM was the only curriculum that had students construct a protractor. Extensions to the measurement of angles were richest in KM, but again, that might be attributed to the block structure that tends to connect mathematical concepts because of the nature of its structure.

Textbooks cannot account for all or even most of the differences found in teacher knowledge and/or student knowledge of fundamental mathematics. But it is clear from our analyses of *Korean Mathematics* in terms of the use of inverse relations of operations, the reliance on base 10 place value, the use of multiple models for concept development, and how concepts build on themselves as well as the development of one concept that textbooks do (or can) make a difference.

REFERENCES

Everyday Mathematics, Grades 1-6. (2002). The University of Chicago School Mathematics Project. Chicago: Wright Group/McGraw Hill.

Fuson, K. C., & Kwon, Y. (1992a). Korean children's single-digit addition and subtraction: Numbers structured by ten. *Journal for Research in mathematics Education, 23*, 148–165.

Fuson, K. C., & Kwon, Y. (1992b). Korean children's understanding of multi-digit addition and subtraction. *Child Development, 63*(2), 491–506.

Grow-Maienza, J. (2002). *Conceptualization of NCTM recommended constructs in Korean textbook materials* (Final report to the NSF for ESIE SGER Award #0086580). Kirksville, MO: Truman State University. Retrieved June 18, 2008, http://eisenhowermathematics.truman.edu/

Grow-Maienza, J., & Beal, S. (2003). *Korean primary mathematics: Block learning and conceptualization of the constructs.* Paper presented at research presession of NCTM annual meeting, San Antonio.

Grow-Maienza, J., & Beal, S. (2005). *What we can learn from Asian mathematics textbooks.* Paper presented at the research presession of the National Council of Teachers of Mathematics annual meeting, Anaheim, CA.

Grow-Maienza, J., Hahn, D. D., & Joo, C. A. (2001). Mathematics instruction in Korean primary schools: Structure, process, and linquistic analysis of questioning. *Journal of Educational Psychology, 93*(2), 363–376.

Houghton Mifflin Company. (2000). *Houghton Mifflin math steps.* Boston: Author.

Korean Mathematics. (2001). Edited by J. Grow-Maienza; translated by S. C. Nugent from *Arithmetic: Grades 1-6* with permission of Korean Ministry of Education. Kirksville, MO: Truman State University.

Li, Y. (2005). *Conceptualizing and organizing content for teaching and learning in Chinese and Singapore mathematics textbooks: The case of division of fractions.* Paper presented at the research presession of the National Council of Teachers of Mathematics annual meeting, Anaheim, CA.

Mayer, R. E., Sims, V., & Tajika, H. (1995). A comparison of how textbooks teach mathematical problem solving in Japan and the United States. *American Educational Research Journal, 32*(2), 443–460.

U.S. Department of Education: National Center for Education Statistics. (1996) *Pursuing excellence, NCES 97-198.*Washington, DC: U.S. Government Printing Office.

Wang, J. (2002). *Beginning teaching mathematics in middle schools: Forms and substance of Chinese teachers' instructional discourses.* Paper presented at the annual conference of the Comparative and International Education Society, Orlando, FL.

Wang, J., & Lin, E. (2005). Comparative studies on U.S. and Chinese mathematics learning and the implications for standards-based mathematics teaching reform. *Educational Researcher, 35*(5), 3–13

Watanabe, T. (2001, April). Content and organization of teacher's manuals: An analysis of Japanese elementary mathematics teacher's manuals. *School Science and Mathematics, 101,* 194–204.

CHAPTER 15

COMPARING ELEMENTARY MATHEMATICS CURRICULA OF KOREA AND THE UNITED STATES

Insook Chung
Saint Mary's College, IN

INTRODUCTION

A Korean mathematics curriculum published by the government and a commercially available U.S. mathematics curriculum for Grades 1 to 5 were compared in terms of book size and volume, overall scope and sequence of mathematics content, and instructional materials and learning games used in the lessons. Teacher manuals, textbooks, and student workbooks were compared to find differences and similarities. Conclusions based on findings are discussed and two suggestions for further research projects are given.

International comparative research projects such as the Third International Mathematics and Science Studies (TIMSS) of 1999 and 2003 reported that Asian students achieved high scores in the subject of mathematics (Gonzales et al., 2000; Gonzales et al., 2004). These reports gener-

Mathematics Curriculum in Pacific Rim Coutries—China, Japan, Korea, and Singapore: Proceedings of a Conference, pp. 255–263

ated increasing interest among educators, researchers, and educational policymakers in examining mathematics curricula from the high achieving countries to seek more effective teaching methods that could improve students' learning and achievement in mathematics. As a result of this movement, various research studies were conducted that revealed some of the curricular issues and critical differences between the mathematics curricula of the United States and Japan (e.g, Saminy & Liu, 1997; Stigler, Gonzales, Kawanaka, Knoll, & Serrano, 1999; Watanabe, 2001), the United States and Singapore (e.g., Seng, 2000; Seng & Thirumurthy, 1999), and the United States and China (e.g., Li, 2000; Yong, 2005). However, few research studies focused on the Korean curriculum even though Korean fourth graders ranked first and eighth graders achieved the second highest score in the TIMSS in 1999. In 2003, the TIMSS reported that Korean eighth-grade students obtained the second highest scores in mathematics. Given these findings and a lack of research in this area, this researcher was motivated to conduct a comparative study of the mathematics curricula of Korea and the United States.

In order to understand and compare the content of the Korean and U.S. mathematics curricula, teacher manuals and textbooks were examined. Analyzing these materials has been commonly done in other educational research studies because they address nationally recognized content and curriculum standards, and they provide classroom teachers with a blueprint for content coverage and an instructional sequence for teaching mathematics in both countries (Li, 2000; National Science Board, 2004). The purpose of this study was to compare the scope and sequence of the mathematics content and instructional materials used for lessons in the elementary curricula of Korea and the United States.

METHODS

Materials

Mathematics teacher manuals, textbooks, and student workbooks were used to compare mathematics curricula from Korea and the United States. The Korean curriculum published under the supervision of the Korean 7th curriculum revision committee (2000/2001) appointed by the Ministry of Education (Ministry of Education and Human Resources Development, 2000a, 2000b) was compared to a U.S. curriculum, *Everyday Mathematics* (The University of Chicago School Mathematics Project) published by Wright Group/McGraw Hill in 2001 (hereafter called *EM*).

Mathematics Content

Overall scope and sequence of mathematics curricula for elementary grades (1-5) in Korea and the United States were analyzed by comparing topics contained in each unit of the teacher manuals and student textbooks. The topics of all of the lessons in the teacher manual and textbooks were analyzed using the content categories of the Standards and Expectations from *Principles and Standards for School Mathematics* (National Council of Teachers of Mathematics, 2000). The first encounter with each topic in the teacher manual or textbook was compared by grade level.

Instructional Materials and Learning Games

Instructional materials and learning games recommended in the teacher manuals were examined in terms of quantity of items and the variety of materials and games used to instruct students were analyzed.

FINDINGS

In the Korean mathematics curriculum, the teacher manuals are illustrated page by page and have exactly the same instructional sequence as the student textbooks. There is a mathematics textbook and student workbook for each semester. The Korean teacher manual for Grade 1 has 503 pages while the *EM* teacher manual has 858 pages (71% more than Korea's) and each page is one inch wider and longer. In the *EM* curriculum there is a *student reference book*, a *student journal*, and two *math master* books (one for each semester) for each grade from Grade 3 through Grade 5.

The Korean mathematics textbook has more units than the *EM* textbook. There are 16 units for Korean first graders while the *EM* first graders have 10 units per school year. Lessons and instructions in the Korean curriculum are sequential and difficulties of the content are gradually increased. Subject matter is treated thoroughly and in depth.

In the following three tables, certain quantitative comparisons are presented between the two curricula. First, Table 15.1 compares the time frame in which mathematics topics are first introduced. Second, Table 15.2 compares the types and numbers of recommended instructional materials. Third, Table 15.3 compares the learning games included in the teacher manuals.

Table 15.1. Comparison of First Introduction to Mathematics Topics

NCTM Standard	Category	Korea	EM	Category	Korea	EM
Number and operations	Zero	1-A-1	Not specifically addressed	Place value	1-A-7	1-A-5
	2-digit addition & subtraction	2-A-2	2-A-4	Multiplication	2-A-8	2-A-6
	Division	3-A-4	2-A-6	Fractions	3-A-7	1-B-3
	Decimals	3-B-6	2-B-4	Estimation	4-B-6	2-A-4
	Percent	6-A-6	4-B-3	Ratio	6-A-6	5-B-2
	Rate	6-A-6	4-B-6	Calculator	Not included	1-A-3
	Integers	Not included	5-B-1			
Algebra	Patterns (geometric)	1-A-3	1-B-2	Patterns (Numeric)	1-B-1	1-A-3
	Algebraic symbols	1-B-3	1-A-2	Representations of data	1-A-8	1-A-1
Geometry	2-dimensional geometric shapes	1-B-2	1-B-2	3-Dimensional geometric shapes	1-A-3	1-B-2
	Congruence & similarity	5-A-2	3-A-6	Motions	2-A-3	4-B-4
	Build geometric shapes (2-D)	1-A-3	1-B-2	Build geometric shapes (3-D)	2-B-3	2-A-5
Measurement	Length	1-A-6	1-A-4	Weight	4-A-5	2-B-3
	Capacity	3-B-5	5-B-3	Volume	6-A-5	2-B-3
	Perimeter	5-A-6	2-B-3	Area	5-A-6	2-B-3
	Time (Analog)	1-B-5	1-A-4	Money	2-A-1	1-A-2
	Temperature (weather)	4-B-7	1-A-1	Unit conversions	2-B-5	2-B-3
Data analysis & probability	Collecting data	1-A-8	1-A-1	Representing/ Displaying Data	2-B-6	2-A-3
	Probability	6-B-6	4-B-1			

Note: The first number of the triplet represents grade levels, the A/B indicates first or second semester, and the last number of each triplet shows the sequence of unit in the teacher manual and textbook. (Thus the data triplets create a lexicographic ordering.)

**Table 15.2.   Comparison of Teacher Manual
Recommended Instructional Materials**

Grade	Korea	United States
1-A	12 (Counter, number card)	26 (Pattern block, penny)
1-B	17 (Base-10 block, straw)	27 (Dominoes, geo-board)
2-A	20 (Paper cup, fruit)	29 (Craft stick, thermometer)
2-B	15 (Button, sticker)	31 (Attribute Block, books)
3-A	19 (Colored paper, candy)	18 (Fact triangle, dictionary)
3-B	25 (mirror, soda bottle)	23 (Recycled items, job chart)
4-A	15 (Compass, pan balance)	17 (Compass, straw, ruler)
4-B	16 (Protractor, grid paper)	23 (Butcher paper, beans)
5-A	19 (Fraction card, box)	21 (Arrays Museum photos, envelope)
5-B	13 (number card, 2 mirrors)	17 (Stock-market pages)
Total	**171** different items	**232** different items

Note: Number of items needed for the instructional activities can vary and thus are approximate.

**Table 15.3.   Comparison of Learning Games
Included in the Teacher Manual**

Grade	Korea	United States
1	15 (Number card/die game)	18 (Addition top-it, difference game)
2	11 (Addition/subtraction number card game)	21 (Array bingo, digit game)
3	12 (Division board game with a die)	14 (Angle race, beat the calculator)
4	11 (Fraction/decimal board game with a die)	23 (Credit/debits game, division dash)
5	14 (Fraction multiplication number card game	33 (Algebra election, factor captor)
Total	**63** games	**109** games

DISCUSSION

The *EM* textbooks and teacher manuals were larger and thicker than their Korean counterparts, and this finding is congruent with what previous research reported (Valverde & Schmidt, 1997/98). This may make U.S. students feel overwhelmed and challenged, especially those students in the primary grades. U.S. teachers may have difficulties addressing all the content within the semester or school year. The Korean teacher's manual illustrates the instructional sequence in a very clear and systematic way so if the teacher follows the lesson plan exactly, even if they are not well trained, basic content can still be covered for each grade level. The Korean education system has been centralized by the government so all public schools have uniform syllabi and textbooks. This guarantees that Korean students from any school district learn relatively the same content at the same rate. However, each school district in the U.S. uses a curriculum selected by a different group, so it is extremely difficult for classroom teachers to maintain uniformity nationally regarding curriculum and syllabi for the subject. In other words, alignment between two districts in content and sequence would be unlikely, thus students who move from district to district may either learn some topics twice or not at all.

In this study, both curricula show a similar scope and sequence of mathematics content. However, the content in the Korean curriculum was clearly sequenced and gradually expanded and increased in difficulty while the *EM* curriculum continuously revisited and spiraled the topics, as earlier researchers claimed (Gill & McPike, 1995; Schmidt, McKnight, & Raizen, 1995; Watanabe, 2001).

In terms of the content, the Korean curriculum introduced about a year later than that of the *EM* the topics of division, fractions, decimals, estimation, algebraic symbols, money, and ratio. The *EM* curriculum included about 2 years earlier than the Korea curriculum the topics of estimation, congruence/similarity, weight, perimeter, area, volume, probability, percent, and rate. Temperature was introduced 3 years earlier in the *EM* curriculum. The Korean curriculum did not encourage the use of calculator at all while *EM* schools used it from the first semester of Grade 1. Integers were not included in the Korean elementary school curriculum. The Korean curriculum introduced motions in geometry 2 years earlier than did the *EM* curriculum. The *EM* curriculum did not specifically introduce zero, while Korea's dealt with it after teaching the number five.

The *EM* curriculum recommended using 232 manipulative items (36% more) in the Grades 1-5 teacher manuals while Korea's curriculum suggested 171 items. The Korean schools predominantly utilized base-10 blocks, number cards/dice, colored papers, black/white counters throughout lessons while *EM* schools used pattern blocks, coins, calculators, dominoes, base-10 blocks, and attribute blocks. The *EM* curriculum included 13 children's storybooks in the instructional materials and integrated literature into mathematics lessons.

The *EM* curriculum utilized a total of 109 learning games for Grades 1-5, which is 70% more games than Korea's curriculum. The Korean curriculum employed 63 games for the instructional activities. While the Korean curriculum predominantly used card or dice games at the symbolic/abstract level, the *EM* curriculum used conventional games with authentic names of the games that might motivate and challenge students to play them. The *EM* curriculum includes more learner-centered materials in the teacher manuals, student journals, and math masters that are hands-on and concrete and authenticated with various learning games and projects using students' real life materials. However, if U.S. teachers do not examine the sequence of the lessons and learning activities compared with the grounding information on why certain activities or games are employed for the topic, helping students understand the content will be very difficult. Many students will have fun with the hands-on activity, although they would not grasp the concept and connect it to skills. The Korean curriculum also contains hands-on learning activities in each unit. The directions of the learning activities and games throughout Grades 1 to 5 are very similar using number cards or dice so students become very familiar with sequence and directions of the learning activities and games. This teacher-friendly format allows Korean teachers to save instructional time in explaining the directions, but Korean students may not be motivated and challenged with the games and activities because of the lack of variety of the game materials and format.

Previous research study (Li, 2000) claimed that Asian curricula contained more difficult and advanced topics than in the United States, but this was not evident in this study of the Korean and *EM* curricula. This issue was discussed in the report of the Korean 7th curriculum revision committee for elementary mathematics. It indicates that Korean elementary mathematics content was reduced by 30% compared to the earlier version (Paik, 2004). Therefore, it is consistent with previous findings that claim mathematics curricula in the United States usually cover more content areas than other countries (Gonzales et al., 2000).

CONCLUSIONS

The results of this study suggest that the Korean mathematics curriculum has an instructional sequence and scope of content arranged in a very clear and structured manner so that classroom teachers are easily able to implement the lessons by exactly following the sequence of the instructional methods. The *EM* curriculum contains a variety of instructional methods using concrete and hands-on learning activities but this may make classroom teachers overwhelmed with the time required to learn the materials and prepare the lessons. This means *EM* teachers may need more training courses to learn the curriculum, especially when a new curriculum is adapted by the school district.

As current learning theory supports using concrete materials, hands-on activities, and interdisciplinary integration of learning activities for students' meaningful mathematics experiences, both curricula included real-life objects and stories from students' actual life situations. The *EM* curriculum utilized more authentic instructional materials, conventional games, nonroutine problem-solving activities, and children's literature to reflect current learning theories on the lessons. However, this curriculum lacks specific focus on content and skills, while the Korean curriculum is extremely sequential and structured, focusing on the scope and sequence of the content. If a curriculum can be developed that is somewhat more focused than the *EM* curriculum and includes more authentic problem-solving activities with interdisciplinary integration, this will be an ideal curriculum for facilitating students' mathematical understanding and improving mathematical achievement scores.

SUGGESTIONS FOR FURTHER RESEARCH

The following research studies should be conducted to implement the results of this study:

1. A study comparing how each topic has been introduced and advanced in the curriculum to help children build conceptual knowledge and connect it to procedural knowledge.

2. A study comparing classroom teachers' instructional sessions to see how the teacher's manual is used and how instructional sequences and techniques are practiced in both countries' actual classrooms.

REFERENCES

Gill, A., & McPike, A. (1995). What we can learn from Japanese teacher's manuals. *American Educator, 19*(1), 14–15.

Gonzales, P., Guzmán, J. C., Partelow, L., Pahlke, E., Jocelyn, L., Kastberg, D., et al. (2004). *Highlights from the Trends in International Mathematics and Science Study (TIMSS) 2003.* Washington, DC: U.S. Government Printing Office

Gonzales, P., Calsyn, C., Jocelyn, L., Mak, K., Kastberg, D., Arafeh, S., et al. (2000). *Highlights from the third international mathematics and science study-repeat.* Washington, DC: U.S. Government Printing Office

Li, Y. (2000). A comparison of problems that follow selected content presentations in American and Chinese mathematics textbooks. *Journal for Research in Mathematics Education, 31*(2), 234–241.

Ministry of Education and Human Resources Development. (2000a). *Mathematics textbooks: Grade 1-5*. Seoul, Korea: Daehan Textbooks.

Ministry of Education and Human Resources Development. (2000b). *Mathematics teacher's manuals: Grade 1-5*. Seoul, Korea: Daehan Textbooks.

National Council of Teachers of Mathematics. (2000). *Principles and standards for school mathematics*. Reston, VA: Author.

National Science Board. (2004). *Science and Engineering Indicators 2004* (2 volumes). Arlington, VA: National Science Foundation.

Paik, S. (2004). *School mathematics curriculum of Korea.* Paper presented at the Tenth International Congress on Mathematics Education (ICME 10), Copenhagen, Denmark.

Saminy, K. K., & Liu, J. (1997). A comparative study of selected U.S. and Japanese first-grade mathematics textbooks. *Focus on Learning Problems in Mathematics, 19*(2), 1–13.

Schmidt, W. H., McKnight, C. C., & Raizen, S. A. (1995). *A splintered vision: An investigation of U.S. science and mathematics executive summary.* Retrieved October 25, 2005, from Michigan State University, U.S. National Research Center for the Third International Mathematics and Science Study Web site: http://ustimss.msu.edu/splintrd.htm.

Seng, S. H. (2000). *Teaching and learning primary mathematics in Singapore.* Baltimore: The International Conference of the Association for Childhood Education International. (ERIC Document Reproduction Service No. ED439812)

Seng, S. H., & Thirumurthy, V. (1999). *Mathematics curriculum in India and Singapore.* San Antonio, TX: The International Conference of the Association for Childhood Education International. (ERIC Document Reproduction Service No. ED429704)

Stigler, J., Gonzales, P., Kawanaka, T., Knoll, S., & Serrano, A. (1999). *The TIMSS videotape classroom study: Methods and findings from an exploratory research project on eighth-grade mathematics instruction in Germany, Japan, and the United States* (NCES 1999-074). Washington, DC: U. S. Government Printing Office.

The University of Chicago School Mathematics Project. (2001). *Everyday mathematics: Grades 1–5.* Chicago: Wright Group/McGraw-Hill.

Valverde, G. A., & Schmidt, W. H. (1997/98). Refocusing U.S. math and science education. *Issues in Science & Technology, 14*(2), 60–66.

Watanabe, T. (2001). Content and organization of teacher's manuals: An analysis of Japanese elementary mathematics teacher's manuals. *School Science and Mathematics, 101*(4), 194–205.

Yong, Z. (2005). Increasing math and science achievement: Best and worst of the East and West. *Phi Delta Kappan, 87*(3), 219–222.

CHAPTER 16

SINGAPORE MATH

Can It Help Close the U.S. Mathematics Learning Gap?

Alan Ginsburg
U.S. Department of Education

Steven Leinwand
American Institutes for Research

On the 2003 Trends in International Mathematics and Science Study (TIMSS; Mullis, Martin, Gonzalez, & Chrostowski 2004), fourth- and eighth-grade students from Singapore achieved the top, average scores in mathematics. This paper compares the Singapore and U.S. primary mathematics system to explore what the United States can learn from the Singapore system of elementary mathematics education that may help improve the mathematics performance of U.S. students. It also identifies a few areas where the U.S. mathematics system may be preferred to Singapore's system.

Before proceeding with the Singapore-U.S. comparisons, we need to address three misimpressions about the Singapore primary mathematics

Mathematics Curriculum in Pacific Rim Coutries—China, Japan, Korea, and Singapore:
Proceedings of a Conference, pp. 265–281

system. First, many experts (Committee on Prospering in the Global Economy of the 21st Century: An Agenda for American Science and Technology, National Academy of Sciences, National Academy of Engineering, Institute of Medicine, 2005) believe that U.S. primary students already perform well on international mathematics assessments and that the focus of U.S. mathematics reform should be on the upper grades. This common perception is based on published reports showing U.S. students scoring above the international average on Grade 4 TIMSS, but falling considerably below the average at age 15 on the Organization for Economic Cooperation and Development's Program for International Student Assessment (PISA; National Center for Education Statistics, 2005). But many industrialized European countries that out score the United States on PISA do not participate in TIMSS. When the United States is compared against the 12 industrialized countries that participate in both TIMSS and PISA, its rank is consistently mediocre on both assessments—eighth on Grade 4 TIMSS and ninth on Grade 8 TIMSS and PISA, age 15. In general, countries that score well at Grade 4 also do so at age 15, so establishing a sound foundation in early mathematics concepts is critical for later learning of more complicated mathematics concepts (Ginsburg, Cooke, Leinwand, Noell, & Pollock, 2005).

Second, it is argued that Singapore's student population is not as diverse as the U.S. student population. But Singapore's population is far from homogenous; almost a quarter of its students are Malay or Indian, along with its majority Chinese population. Further, Singapore's minority students far outperform U.S. minority students. To illustrate, four out of five Singapore's Malaysian students score in the top half of the TIMSS-2003 at Grade 8, whereas a majority of U.S. Black students score in the bottom quarter of the TIMSS international student rankings (Ministry of Education, Singapore, 2001; National Center for Education Statistics, 2000).[1]

Third, it is argued that Singapore's small population of about 4 million allows the development of a strong centralized curriculum not possible to replicate in the highly decentralized U.S. education system. True, but U.S. states under the federal No Child Left Behind legislation must have in place challenging mathematics standards and uniform assessments in at least Grades 3–8, and states could adopt a curriculum like Singapore's (U.S. Congress, 2002).

Given Singapore's success, it is worthwhile to carefully analyze Singapore's mathematics education system and how it compares with the U.S. education system as the United States seeks to achieve the goals and mandates of No Child Left Behind. The comparisons of the two systems are made in the areas of frameworks, textbooks, assessments, and teachers, with the key findings described below.

FRAMEWORKS

- *The U.S. lacks Singapore's uniform and coherent national mathematics curriculum that guides its mathematics program.*

Singapore has a well-defined, specific national mathematics framework. By contrast, the United States has no official national mathematics framework. In fact, the U.S. Department of Education is prohibited by law from specifying a national curriculum (U.S. Congress, 1979). Instead, the National Council of Teachers of Mathematics (NCTM, 2000) standards are the closest U.S. approximation to national standards and are the basis for many state frameworks.

However, the NCTM framework lacks Singapore's highly logical and specific mathematics framework. Singapore specifies grade-specific content, but NCTM specifies imprecise content only within broad grade bands (K–2, 3–5). Singapore adopts a spiral approach that builds outward on prior content; the U.S. NCTM uses a spiral approach that to a significant degree repeats content across grades.

The treatment of multiplication in Grades 1 and 2 illustrates the differences between the Singapore and NCTM framework approaches. Singapore's Grade 1 standards are specific and bounded—multiplication of numbers with a product not greater than 40 and division of numbers with a result not greater than 20. Grade 2 expands multiplication and division within the 2, 3, 4, 5, and 10 times tables.

In contrast with Singapore's specific and cumulative approach, NCTM covers multiplication and division in Grades 1 and 2 within a broad K–2 grade band and within the general category of the "meaning of operations" rather than multiplication and division. Students are to "understand situations that entail multiplication and division, such as equal groupings of objects and sharing equally." These NCTM statements are too broad and unspecific to provide the direction necessary to guide what mathematics states and school systems should teach at specific grades.

- *State frameworks differ greatly; some state frameworks are nearly as specific and focused as Singapore's but others are poorly organized and lack direction.*

The U.S. state mathematics frameworks differ considerably in the amount and nature of the mathematics content at each grade. Table 16.1 illustrates the differences between the mathematics frameworks of seven states and that of Singapore. Three of the seven states, California, North Carolina, and Texas, expose students to nearly the same amount of content as Singapore. Florida, in contrast, specifies about twice the number

Table 16.1. Comparison of Content Exposure for Singapore and Selected U.S. States Mathematics Standards: Grades 1–6

| | | Avg. No. of Topics/Grade | | Avg. No of Grades/Topic | | | Avg. No. of Outcomes/Grade | |
	Total No. of Topics	No.	Ratio to Singapore	No.	Ratio to Singapore	Total No. of Outcomes	No.	Ratio to Singapore
Singapore	40	15	—	2.3	—	232	39	—
California	42	20	1.3	2.9	1.3	305	51	1.3
Florida	54	39	2.6	4.2	1.8	640	107	2.7
Maryland	46	29	1.9	3.8	1.7	415	69	1.8
New Jersey	50	28	1.9	3.4	1.5	336	56	1.4
North Carolina	41	18	1.2	2.6	1.1	217	36	.9
Ohio	48	26	1.7	3.3	1.4	370	62	1.6
Texas	40	19	1.3	2.8	1.2	265	44	1.1

Source: Ministry of Education, Singapore (2001); States are from standards available in 2005 from individual state Web sites.

of mathematics topics per grade; teaches each topic twice as long; and also expects students to learn twice the number of outcomes per grade. Maryland, New Jersey, and Ohio frameworks fall in between in content exposure. With so much more mathematics content to cover in each grade, many U.S. state frameworks are not able to treat topics with depth. Therefore, they repeat the same topics over many more grades than does Singapore, and the mathematics curriculum does not progress at as fast a pace.

The treatment of fractions in Grade 4 in Singapore and in Florida illustrates Singapore's greater focus and depth of presentation. Singapore specifies that students should be able to do the following:

- Add and subtract like fractions; related fractions (Denominators of given fractions should not exceed 12; exclude sums involving more than two different denominators.).
- Calculate the product of a proper fraction and a whole number.
- Express an improper fraction as a mixed number, and vice versa (Include expressing an improper fraction/mixed number in its simplest form).
- Solve up to two-step word problems involving fractions. (Include using unitary method to find the "whole" given a fractional part.

Exclude question such as "Express the number of girls as a fraction of the number of boys" as it will be dealt with under the topic "Ratio.")

Contrast Singapore's clear focus on the mathematics of fractions with Florida's Grade 4 specifications involving fractions:

- Compares and orders commonly used fractions and decimals to hundredths using concrete materials, drawings, and numerals.
- Locates whole numbers, fractions, mixed numbers, and decimals on a number line.
- Translates problem situations into diagrams and models using whole numbers, fractions, mixed numbers, and decimals to hundredths including money notation.
- Uses concrete materials to model equivalent forms of whole numbers, fractions, and decimals.
- Knows that two numbers in different forms are equivalent or non-equivalent, using whole numbers, decimals, fractions, and mixed numbers.
- Explains and demonstrates the addition and subtraction of common fractions using concrete materials, drawings, story problems, and algorithms.
- Explains and demonstrates the addition and subtraction of decimals (to hundredths) using concrete materials, drawings, story problems, and algorithms.
- Predicts the relative size of solutions in addition and subtraction of common fractions.
- Uses problem-solving strategies to determine the operation(s) needed to solve one- and two-step problems involving addition and subtraction of fractions.

Florida's framework mixes the learning of whole numbers, decimals, and mixed numbers with fractions. Florida's descriptions also intersperse what students need to know about fractions with very general prescriptions about teaching fractions using concrete materials and drawings. In addition, Florida's descriptions lack the precision of Singapore's statements, such as "calculate the product of a proper fraction and a whole number."

- ***Singapore recognizes that some students may have more difficulty in mathematics and provides them with an alternative framework; the U.S. frameworks make no such provisions.***

**Table 16.2.   Comparison of Textbook Organization by
Number of Topics, Lesson Length, and
Pages Devoted to Development and Exercises: Grade 3**

	Total Pages	No. of Chapters	No. of Lessons	Average Pages per Lesson	Pages of Development	Pages of Exercises	Other Pages
Singapore	496	14	42	12	150 (30%)	234 (47%)	112 (23%)
Scott-Foresman	729	32	164	4	187 (26%)	217 (30%)	325 (45%)
*Everyday Math**	250	11	120	2	145 (58%)	105 (42%)	—

Sources: Ministry of Education, Singapore (2001), Scott-Foresman (2004), and *Everyday Mathematics* (2001).
Note: *Includes student journals, activity book, math in my world and math masters.

Singapore has developed an alternative mathematics framework for lower-performing students in the upper primary grades that covers all the mathematics topics in the regular framework, but at a slower pace and with greater repetition. Singapore also provides its slower students with extra help from well-trained teachers. NCTM and the states we examined provide no alternative framework for lower-performing mathematics students. Moreover, such students are often tracked into slower mathematics courses, but unlike in Singapore, these students are seldom taught all the required mathematics material. Evaluations have shown that they frequently receive their extra help from teacher's aides who lack college degrees and/or background in mathematics.

TEXTBOOKS

- ***Singapore textbooks' in-depth treatment of mathematical topics build deep understanding of mathematical concepts; traditional U.S. textbooks rarely get beyond definitions and formulas, developing primarily students' mechanical ability to apply mathematical concepts.***

Singapore textbook publishers know the mathematics that students are expected to learn grade by grade. U.S. textbook publishers know the textbook content for some states with well-defined frameworks, but not for other states where frameworks specify mathematical content only in a general way. Even when states have well-defined mathematics topics in each grade, the same topics are taught over different grades in different states. As an example, students are expected to gain fluency in the addition of fractions by Grade 4 in one state; Grade 5 in 15 states; Grade 6 in 20 states; and Grade 7 in six states (Reys, 2006).

U.S. textbook publishers seek to maximize a textbook's market share across states with different frameworks by including lessons on many more topics in the same textbook than is the case in Singapore. To quantify this difference in textbook topic coverage, we compare the Singapore mathematics textbook and a traditional U.S. textbook we selected, the Scott-Foresman mathematics series (Scott Foresman-Addison Wesley, 2004). We also selected a nontraditional U.S. mathematics textbook, *Everyday Math,* which was developed as part of the National Science Foundation textbook development effort (Wright Group/McGraw-Hill, 2001).

These three textbooks are compared in their treatment of the Grade 3 topics in Table 16.2. Comparisons for other grades yield similar results. The Singapore textbook covers half the number of topics as the two U.S. textbooks. Moreover, there are 12-page lessons on average in Singapore, compared with Scott-Foresman's short four-page lessons and *Everyday Math's* two pages of student materials per lesson. Although the U.S. traditional textbook has almost double the pages of Singapore's textbook, about 45% of the pages do not cover either concept development or exercises, far more than for either the Singapore textbook or *Everyday Math.*

The lengthier lessons allow the Singapore textbook to have rich and cumulative topic development. We illustrate the richness of the Singapore development with the pie chart problem in Figure 16.1a. In order to calculate the money collected by the different game stalls, students need to understand that the size of the pie slices are in proportion to the size of the angles in a circle and draw on their knowledge of right angles and the sum of angles of a straight line. Solving the pie chart example from the Scott-Foresman textbook (Figure 16.1b) does not requiring using properties of angles and requires knowing only that the sum of the pieces of the pie have to equal the total, or 100%.

SINGAPORE'S CONCRETE-PICTORIAL-ABSTRACT APPROACH

Students begin to grasp the power of mathematics when they are able to apply the essential core mathematical principles to solve a rich variety of mathematics problems. But to grasp mathematical ideas, students must understand and apply mathematical abstractions, yet abstract mathematics is difficult for many students to grasp. The Singapore textbook fea-

The pie chart represents the amount of money collected by various stalls at a funfair.

(a) What fraction of the total amount of money was collected by the games stalls?
(b) What was the total amount of money collected by the various stalls?
(c) How much money was collected by the music stalls?
(d) What was the ratio of the money collected by the food stalls to the money collected by the handicraft stalls?

Figure 16.1a.   Singapore's pie chart problem with intermediate unknowns involving angles.

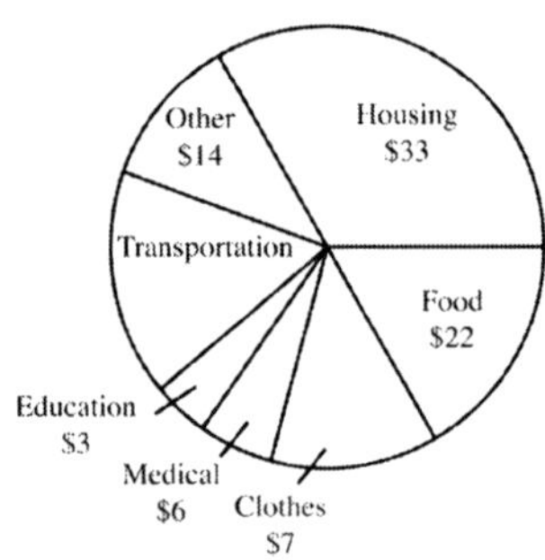

(a) What is the cost of transportation?
(b) For each $100 a parent spends raising a child to age 18, how much more is spent on housing and clothes than on education?
(c) TEST PREP. For each $300 spent, estimate how much is spent for food and clothes
       1) $329   2) $90   3) $29   4) $130
(d) Which costs are about twice as much as the cost of education? Five times as much? Eleven times as much?

Figure 16.1b.   Scott-Foresman pie chart problem with mechanical application of definition of pie chart.

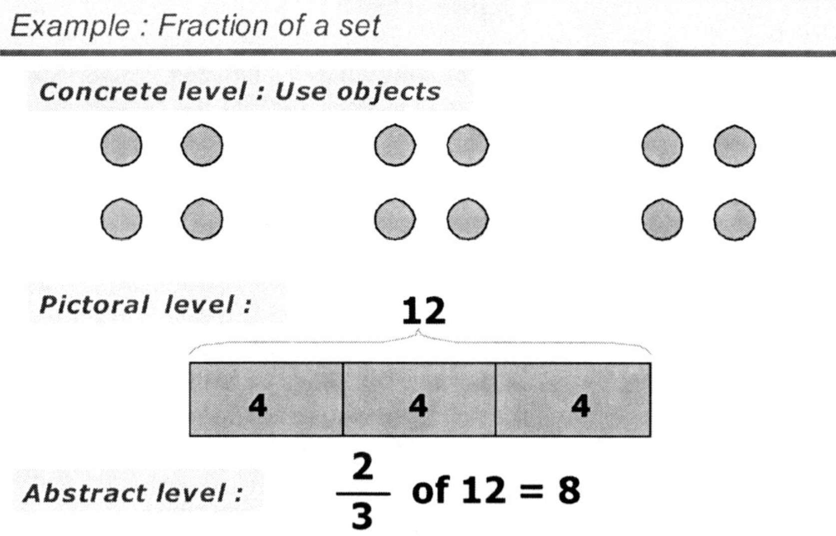

Source: Singapore Primary Math Grade 3.

Figure 16.2.   Singapore's concrete-pictorial-abstract approach provides a bridge for students who have trouble with abstract mathematics concepts.

tures instructional presentations in which a concept is first illustrated concretely, then pictorially, and finally abstractly. The approach is similar in purpose to the manipulatives common in U.S. schools, but Singapore tightly connects its concrete examples with student learning of mathematical ideas, a frequent weakness with the U.S. application of manipulatives.

Figure 16.2 illustrates the concrete-pictorial-abstract approach applied to explaining how $\frac{2}{3}$ of 12 is the same as dividing 12 into thirds (i.e., three equal groups of size 4) and adding together two of the three groups of thirds. The concrete level displays thirds in terms of objects. The pictorial level shows thirds represented by a bar. Finally, the abstract level expresses the fraction of thirds in terms of an equation. Singapore also builds up the pictorial approach to show how to use diagrams to translate complicated word problems into mathematical expressions. Similar consistent presentations of the concrete, pictorial, and abstract approach were not found in the traditional U.S. mathematics textbooks.

ASSESSMENT

- *Singapore has uniform assessments at two transition grades, 6 and 10; the U.S. mandates testing in more grades, but the 50 state assessments of mathematics differ significantly in how they measure student performance.*

Singapore's uniform assessments are developed by the Ministry of Education at Grades 6 and 10. These are high stakes placement exams. The uniform national examination covers the cumulative mathematics content in Singapore's mathematics framework. In the United States, there is no official national assessment. The No Child Left Behind Act (NCLB) does provide for individual state assessments in Grades 3–8 and one high school grade. At the national level, the U.S. Department of Education administers the National Assessment of Educational Progress (NAEP), but it is given to only a sample of each state's students in mathematics in Grades 4 and 8; scores are reported for states, but not for individual students.

In response to NCLB, each state's tests are variable yardsticks on which a student's test results often depend on the state in which he or she resides. One way to quantify the differences in state measures is to compare a state's pass rate of its students on its own test with its pass rate on the uniform national NAEP test for the same grade. South Carolina is a state that has set a stringent state mathematics test, on which only 23% of the state's students achieved proficient levels, below the 30% at least proficient on NAEP's mathematics assessment. Tennessee is a state that established much lower pass standards for students to achieve proficient or above on its own test. Although 87% of Tennessee's Grade 8 students were proficient on Tennessee's own state test, only 21% were proficient on NAEP (Dillion, 2005). Until the United States decides to have state tests match up with a uniform national test like NAEP, these disparities in test measures across states will continue.

Singapore's school accountability approach measures the school's progress of the same students over grades (i.e., value-added); the U.S. school accountability approach in NCLB measures the progress of different students at the same grade over time.

Singapore uses a value-added measure of learning growth. Grade 10 actual scores are compared with Grade 10 predicted scores, where predicted scores are based on Grade 6 student scores. Those schools with actual grade 10 scores above predicted grade 10 scores are rewarded with recognition and funding (Tan, 1995).

NCLB measures school performance by whether school outcomes at each grade demonstrate adequate yearly progress (AYP) with successive cohorts of students at the same grade. States establish annual improvement goals

Table 16.3. Comparison of Assessment Items for Singapore, Selected States, and NAEP (Percent of Items)

	Multiple Choice	Multistep	Finding an Intermediate Unknown
Singapore, Gr. 6	31%	25%	19%
Florida, Gr. 8	52%	12%	8%
New Jersey, Gr. 8	85%	33%	0%
N. Carolina, Gr. 8	100%	5%	5%
Ohio, Gr. 6	74%	17%	4%
Texas, Gr. 8	100%	6%	2%
NAEP*, Gr. 8	65%	21%	8%

Source: State data based on released items; NAEP data based on entire item pool.

for each grade, and schools must meet state progress at each grade. Schools need to continue to demonstrate progress so that by 2014, all students in a school achieve proficient levels. A criticism of this approach is that schools that have more needy students and that start with lower performing students must demonstrate more growth to reach proficiency than schools with higher performing students. Hence, schools are discouraged from taking low-income students, for example, where school choice plans exist. This is not a problem with Singapore's value-added approach in which schools are compared by the amount of improvement. To test the feasibility and benefits of a value-added measure, Secretary of Education Spellings recently announced a pilot that would allow some states to use a Singaporelike, value-added approach on a trial basis (U.S. Department of Education, 2005).

The questions on Singapore's high stakes Grade 6 Primary School Leaving Examination (PSLE) are more challenging than the released items on the U.S. Grade 8 National Assessment of Educational Progress or the Grade 8 state assessments.

The items on the Singapore assessments are more likely to require deeper problem-solving skills than items on the U.S. state assessments or NAEP at Grade 8. Items on Singapore's tests are more likely to require preparing constructed responses, solving for intermediate unknowns, and developing nonroutine solutions than are U.S. test items (see Table 16.3).

TEACHERS

- *Singapore teachers must pass a rigorous entrance exam to be accepted to education school, which they are paid to attend; U.S. teachers most frequently pass the Praxis I and Praxis II elementary*

> *teacher exams, which are no harder than the Singapore Grade 6 exam.*

Singapore targets acceptance to education school from the top third of its graduates by using a rigorous exam as part of the screening process. Singapore's primary education teachers typically have only 2 years of a college education, but during this period they take 12 semester hours of mathematics pedagogy courses covering the content and teaching of all the major topics in the Singapore framework.

U.S. elementary teachers are among the poorest performers of all college students in mathematics, as measured by their SAT scores, which are significantly below the SAT scores of the average college graduate (Gitomer, Latham, & Ziomek, 1999). Education majors take less mathematics than the typical college graduate, 6.3 credit hours compared with 8.3 credit hours (National Center for Education Statistics, 2005). They are able to start with low mathematics scores and take few courses to improve their mathematics because of the low requirements of the Praxis I at the beginning of education school and the Praxis II elementary exams taken for state licensing.

The most common exams that U.S. teachers have to pass are the Praxis I to graduate from education school and the Praxis II to obtain a state license. As the typical Praxis I questions illustrate, the Praxis elementary teacher mathematics questions are all multiple-choice questions, which rarely require examinees to do more than solve one-step, straightforward applications of a definition or concept. Contrast these questions with three illustrative questions at the easy or moderate level from Singapore's Primary School Leaving Exam at Grade 6 (see Figure 16.3). The Singapore Ministry of Education designated easy questions as ones that all students would be expected to answer correctly and moderate questions as ones that average students would be expected to answer correctly. The Singapore questions require students to undertake multiple step solutions and demonstrate an understanding of concepts in nonroutine situations. Toughening the assessments that elementary teachers pass is one of the surest way to obtain teachers who understand core mathematics ideas.

AREAS OF STRENGTHS IN THE U.S. MATHEMATICS SYSTEM COMPARED WITH SINGAPORE'S SYSTEM

Although the U.S. mathematics program is weaker than Singapore's in most respects, the U.S. system is stronger than Singapore's in some areas.

Typical U.S. PRAXIS I Teacher Elementary Exam Questions (source: http://ftp.ets.org/pub/tandl/0730.pdf))

1. Which of the following is equal to a quarter of a million?

   (A) 40,000   (B) 250,000   (C) 2,500,000   (D) 1/4,000,000   (E) 4/1,000,000

2. Which of the following fractions is least?

   (A) 11/10   (B) 99/100   (C) 25/24   (D) 3/2   (E) 501/500

3. Which of the sales commissions shown below is greatest?

   (A) 1% of $1,000   (B) 10% of $200   (C) 12.5% of $100   (D) 15% of $100   (E) 25% of $40

4. For a certain board game, two dice are thrown to determine the number of spaces to move. One player throws the two dice and the same number comes up on each of the dice. What is the probability that the sum of the two numbers is 9?

   (A) 0   (B) 1/6   (C) 2/9   (D) 1/2   (E) 1/5

Typical Singapore Grade 6 Primary School Leaving Exam
1. Singapore Grade 6 (easy) Question: Global pours an equal amount of water into two empty P and Q tanks shown below.

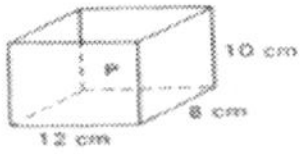
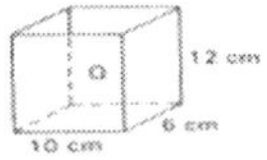

If tank P is half-filled, what is the height of the water level in Tank Q?

A) 5 cm   B) 6 cm   C) 8 cm   D) 10 cm

2. Singapore Grade 6 (moderate). This figure is made up of two semicircles. Q is the centre of the larger semicircle of radius 8 cm.

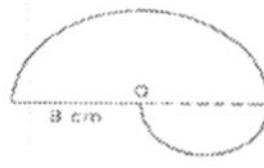

The area of the figure is _____ cm².

   (1) 12π   (2) 40π   (3) 48π   (4) 80π

Singapore 6th grade (moderate). Jim uses 1-cm cubes to build a solid. He starts with a square base using 100 cubes. The second layer of 81 cubes and the third layer of 64 cubes are also arranged in the form of squares. The figure shows the three layers of cubes.

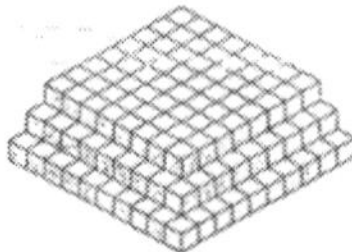

Jim continues building the solid in the same way.
a) How many cubes will there be in the next layer?
b) If the topmost layer consists of 9 cubes, how many layers does the solid have?

Figure 16.3.   Typical questions on U.S. Praxis exam for prospective elementary teachers compared with Singapore Grade 6 Primary School Leaving Exam sample problems (easy or moderate difficulty).

- ***The U.S. framework gives greater emphasis than Singapore's framework does to developing important twenty-first-century mathematical skills such as representation, reasoning, making connections, and communication.***

Singapore's framework emphasizes a very traditional set of school processes. These include understanding concepts; performing processes including arithmetic; and strategic problem solving, such as heuristic strategies for formulating problems (e.g., use a diagram or model, work backward, simplify the problem, look for patterns).

The U.S. NCTM and many state frameworks emphasize twenty-first-century skills rather than the traditional skills of Singapore's focus. Whereas Singapore has a single "process" category for strategic problem-solving skills, NCTM (2000) has three: reasoning and proof, representation, and connections, in addition to problem solving. NCTM makes communication of mathematical ideas a priority, but Singapore does not give it much emphasis and includes communication only as one on a list of

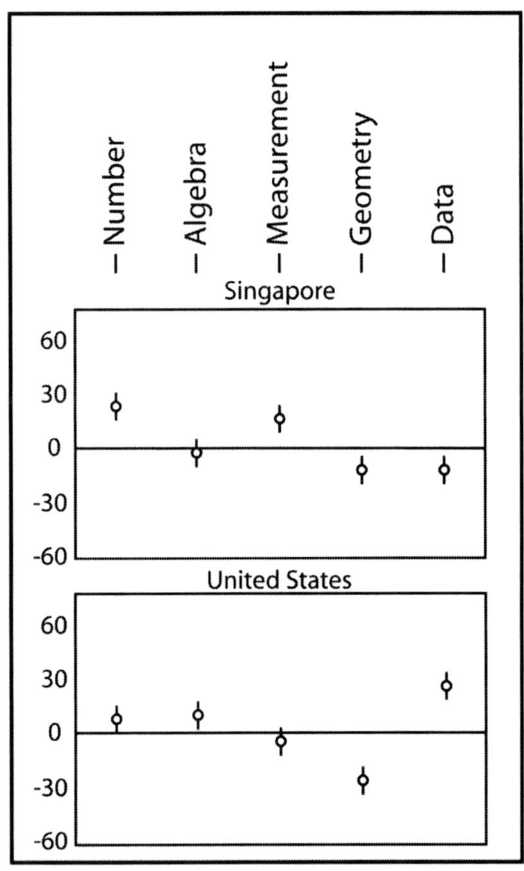

Source: Mullis et al. (2004).

Figure 16.4.   Relative strengths and weaknesses of Singapore and U.S. curriculum, TIMSS-8, 2003.

skills. If U.S. students could grasp twenty-first-century mathematics skills, they could leap frog their international performance, but the NCTM and the state frameworks need to do a better job of integrating them with rigorous mathematics content.

- ***The United States places a greater emphasis on applied mathematics, including statistics, probability, and real-world problem analysis.***

The U.S. mathematics frameworks stress data analysis and probability, whereas the Singapore framework treats statistics in a strictly theoretical way. This difference in content emphasis is apparent in comparing the traditional U.S. textbook, which at Grade 6 has 11% of the lessons about data and statistics, with Singapore's textbook, which has only 1%. Figure 16.4 shows that the relative time devoted to a standard makes a difference as evidenced by the fact that the U.S. scores on TIMSS-8 are relatively strong in data (including statistics) and Singapore's scores are relatively weak in data.

The U.S. frameworks and textbooks also place greater emphasis on mathematical applications. This is clearly seen in the nontraditional textbook, *Everyday Math*, which uses a problem-based learning approach, often involving multistep, real-world mathematics problems as the primary method of instruction. Such applications give students practice in understanding how to apply mathematics in practical ways. Singapore's textbooks would benefit from more real-world applications.

IMPLICATIONS

Singapore's mathematics system is far better designed to truly *leave no child behind* than is the current U.S. mathematics system. Singapore's framework meets NCLB provisions to set "challenging academic standards." Its instruction is "research based." Its teachers are truly "highly qualified," as measured by their knowledge of mathematics, and its assessments are "aligned" with the content frameworks and assess students' mathematics knowledge at both "proficient and advanced levels." The result is that Singapore students, including its minority students, are among the best students in the world. The United States, while retaining its emphasis on statistics, applications, and twenty-first-century skills, would do well to model its mathematics reforms after the superior features of Singapore mathematics.

NOTES

1.   Singapore's tested population in TIMSS does not include children of guest workers as these children do not reside in Singapore and attend school in another country. This is consistent with the treatment of U.S. immigrant workers' children who reside in another country and are excluded from the U.S.-reported TIMSS population.

REFERENCES

Dillion, S. (2005, November 26). Students ace state tests, but earn D's from U.S. *The New York Times*. Retrieved June 19, 2008, from http://www.nytimes.com/2005/11/26/education/26tests.html

Gitomer, D., Latham, A., & Ziomek, R. (1999). *The academic quality of prospective teachers: The impact of admissions and licensure testing*. Princeton, NJ: Educational Testing Service.

Ginsburg, A., Cooke, G., Leinwand, S., Noell, J., & Pollock, E. (2005). *Reassessing U.S. international mathematics performance: New findings from the 2003 TIMSS and PISA*. Washington, DC: American Institutes for Research.

Ginsburg, A., Leinwand, S., Anstrom, T., & Pollock, E. (2005). *What the United States can learn from Singapore's world-class mathematics system (and what Singapore can learn from the United States)*. Washington, DC: American Institutes for Research.

Gonzales, P. A. (2001). *Pursuing excellence: Comparisons of international eighth-grade mathematics and science achievement from a U.S. perspective, 1995 and 1999: Initial findings from the Third International Mathematics and Science Study-Repeat*. Washington, DC: National Center for Education Statistics.

Ministry of Education, Singapore. (2000). *Singapore number one in mathematics again in the Third International Mathematics and Science Study 1999 [TIMSS-R]* (Singapore Ministry of Education Press Release). Singapore: Author. Retrieved June 19, 2008, from http://www.moe.edu.sg/press/2000/pr06122000.htm

Ministry of Education, Singapore. (2001). *Primary mathematics syllabus* (Curriculum Planning and Development Division). Retrieved on June 19, 2008, from http://www.moe.gov/sg/education.syllabuses/sciences/files/maths-primary-2001.pdf

Ministry of Education, Singapore. (2003). *Primary school leaving exam*. Retrieved on June 19, 2008, from http://www.moe.edu.sg/exams/psle/psle.htm

Mullis, I., Martin, M., Gonzalez, E., & Chrostowski, S. (2004). *TIMSS 2003 international mathematics report*. Boston: TIMSS and PIRLS International Study Center, Lynch School of Education, Boston College.

National Academies Committee on Prospering in the Global Economy of the 21st Century (U.S.). (2005). *Rising above the gathering storm: Energizing and employing America for a brighter economic future*. Washington, DC: National Academies Press.

National Center For Education Statistics. (2000). *Pursuing excellence: Comparisons of international eighth-grade mathematics and science achievement from a U.S. perspective, 1995 and 1999*. Washington, DC: Office of Educational Research and Improvement, U.S. Department of Education: NCES 2001–028.

National Council of Teachers of Mathematics. (2000). *Principles and standards for school mathematics*. Reston, VA: National Council of Teachers of Mathematics.

Reys, B. (Ed.). (2006). *The intended mathematics curriculum as represented in state-level curriculum standards: Consensus or confusion?* Charlotte, NC: Information Age.

Scott Foresman-Addison Wesley. (2004). *Mathematics*. Glenview, IL: Pearson Education.

SingaporeMath.com Inc. (2002). *Primary mathematics scope and sequence*. Retrieved July 4, 2008, from http://www.singaporemath.com/Scope_and_Sequence_s/120.html

Tan, Y. K. (1995). Performance indicators for school management (PRISM). In A. Ginsburg & E. Owen (Eds.), *APEC performance measurement systems symposium*. Washington, DC: Education Forum.

The University of Chicago School Mathematics Project. (2001). *Everyday mathematics*. Chicago: Wright Group/McGraw-Hill.

U.S. Congress. (1979). Public Law 96-88.

U.S. Congress. (2002). *No Child Left Behind Act of 2001*. Public Law 107–110 107th Congress.

U.S. Department of Education. (2005, November 18). *Secretary Spellings announces growth model pilot* (Press release). Retrieved June 19, 2008, from http://www.ed.gov/news/pressreleases/2005/11/index.html

CHAPTER 17

SINGAPORE MATH

Perspectives and Experiences of a College Professor

Richard Bisk
Worcester State College

INTRODUCTION

This paper will describe some of my work using Singapore math textbooks (Kong, 2003; Meng & Yoong, 1997) as a professional development tool for improving elementary and middle school teacher understanding of the mathematics they teach. The key characteristics of the K–8 books will also be discussed; particularly those that highlight what U.S. educators might learn from the Singapore approach.

PROFESSIONAL DEVELOPMENT

I started using Singapore math in professional development courses in 2000 as a vehicle to connect teacher knowledge of mathematical content with elementary and middle school student work. The first four courses,

Mathematics Curriculum in Pacific Rim Coutries—China, Japan, Korea, and Singapore:
Proceedings of a Conference, pp. 283–290

taught in 2000, 2001 and 2002, were funded by the Massachusetts Department of Education's (MassDOE) Summer Content Institute program. Three of the courses had an audience of middle school teachers, while the fourth was designed for elementary teachers. The courses were hosted by the North Middlesex Regional School District (NMRSD) and were team taught with three of their middle school teachers.

Although the primary goal was to improve teacher content knowledge, NMRSD also hoped to start a pilot project using the books in middle school classrooms. After the first summer, five teachers used the materials in a total of six classes. The project grew to 18 teachers in 19 classrooms in 2001–2002; and then 49 teachers in 55 classrooms in 2002–2003. NMRSD currently uses the Singapore books as the primary text in all of their K–8 classrooms.

In the summer of 2003, MassDOE funded a larger program involving 49 teachers from nine school districts throughout the state who were to use the Singapore materials in their classrooms the following year. The primary goal of the program shifted to preparing participants to teach using the books. However, we still maintained a focus on improving teacher understanding of the mathematics they teach. For the first time, we had a textbook: *Elementary Mathematics for Teachers* by Thomas Parker and Scott Balderidge (2003). This book used an approach that was consistent with our philosophy; namely improve understanding of elementary mathematics while using the Singapore math texts to make connections to classroom practice.

In the process of teaching these courses, I learned much about the mathematics teachers understanding. We used a pretest/posttest model to measure our success in teaching the mathematics. Often we would use questions that came directly from the Singapore texts. In a 2004 Summer Content Institute for 28 teachers, we included three word problems from the sixth-grade books. The mean score on these problems, out of a possible total of 8 points, was 1.18 on the pretest and 5.57 on the posttest. The good news is that we saw significant growth. However, both the initial and final scores are indicative of the limited mathematical proficiency of many elementary and middle school teachers.

This issue of teacher content knowledge is the biggest challenge we face in improving K–8 mathematics instruction. We would never be satisfied if our third-grade teachers read at the sixth-grade level. However, we have accepted that many operate mathematically at the sixth-grade level. This is not meant to be a criticism of teachers, but rather of some of our teacher-training programs and state departments that license teachers. Many elementary school teachers will readily admit that they do not feel comfortable with mathematics. I believe that teacher-content knowledge is critical and see the Singapore math books as a vehicle for improving it.

How do you teach a mathematical subject when you are not proficient in it? You focus on rules, procedures, and memorization; or on manipulatives, games, and activities that you cannot readily connect to concepts. (I should note that none of these tactics are bad if done appropriately.) When I work with teachers or students, the overriding point that I want to get across is the importance of understanding whatever mathematics they are doing. When you resort to teaching or learning by memorization only, nothing is being taught or learned. The focus on understanding is implicit in the Singapore math books.

In the current "math wars," we often see two positions about elementary mathematics: reform and traditional. Many see these positions as diametrically opposed; traditional focuses on basic skills, while reform emphasizes conceptual understanding. But a student who understands place value should have no difficulty multiplying two 2-digit numbers. I want students to memorize their times tables; but to do the memorization with understanding. When they can't remember what 6×8 equals; they might think: "I know that $5 \times 8 = 40$. So one more 8 is 48." Or they might think: "I know that $3 \times 8 = 24$. To get 6×8, I need to double that result." These are just two of the many areas where basic skills and conceptual understanding support each other. The Singapore books do an excellent job of teaching for understanding *and* emphasizing the importance of basic skills.

THE K–8 SINGAPORE BOOKS

The following list summarizes the key features of the Singapore primary math textbooks.

- Depth emphasized over breadth: Compared to U.S. books, more time is spent on each topic. Fewer topics are covered in a year. There is a greater focus on mastery.

- Problem-solving emphasis: Model drawing diagrams are used to promote understanding of word problems and provide a bridge to algebraic thinking.

- More multistep problems: Problems often require the use of several concepts.

- Mental math: Techniques encourage understanding of mathematical properties and promote numerical fluency.

- Absence of clutter and distraction: Presentation is clean and clear and uses simple, concise explanations.

- Coherent development: Topics are introduced with simple examples and then incrementally developed until more difficult problems are addressed.

- Teacher and parent friendly: Since mathematical content is clear, it is often easier for teachers to plan lessons. Parents can read the books and help children.

- Review of concepts is not explicitly incorporated into the curriculum. Students are expected to have mastered a concept once it has been taught. However previously learned concepts are regularly used to promote retention of understanding.

- A high level of expectation is implicit in the curriculum.

The remainder of this paper will briefly address the first four items on the above list. Each is related to the issue of teaching for understanding.

DEPTH VERSUS BREADTH

When I mention this issue to teachers who are using U.S. books, there is usually a collective nodding of heads. Many tell me that they have to race through their curriculum and do not have time to focus on building conceptual understanding. Table 17.1 is adapted from the American Institute for Research Study: *What the United States Can Learn From Singapore's World-Class Mathematics System* (Ginsburg, Leinwand, Anstrom, & Pollock, 2005). The table compares the Singapore *Primary Mathematics* books with two U.S. texts, the reform oriented *Everyday Math* books and the more traditional Scott-Foresman series (Scott Foresman-Addison Wesley, 2004; Wright Group/McGraw-Hill, 2001). The number of lessons in these books for the given grades ranges from three to six times as many as in the corresponding Singapore materials.

Table 17.1.   Comparison of Singapore Primary Mathematics Book to Everyday Math and Scott-Foresman

	Grade 1		Grade 3		Grade 6	
Textbook	*Number of Lessons*	*Avg. Pages/ Lesson*	*Number of Lessons*	*Avg. Pages/ Lesson*	*Number of Lessons*	*Avg. Pages/ Lesson*
Singapore	34	15	42	12	24	17
Scott-Foresman	157	4	164	4	158	5
Everyday Math	110	2	120	2	113	4

PROBLEM SOLVING EMPHASIS/MORE MULTISTEP PROBLEMS

Problem solving is one of the key components of the Singapore curriculum. This is consistent with the recommendations in the National Council of Mathematics' *Principles and Standards for School Mathematics* (2000) which states that "Problem solving is an integral part of all mathematics learning" (p. 52). The model drawing approach used in the Singapore books takes students from the concrete to the abstract stage via an intermediary pictorial stage. The process provides students with a tool for tackling multi-step problems, where often they would get lost in the details. It also provides an excellent bridge to algebra. Model drawing involves the use of bar diagrams that are broken into units. These units are precursors to variables used in algebra. Students are taught to carefully label the problems with the information provided in the problem and to use a "?" to indicate what they are trying to find. They are also expected to answer the problem in complete sentence form.

Examples from Grades 3–6, shown in Figures 17.1, 17.2, and 17.3, should provide the reader with an initial understanding of this technique. Note that the diagrams contain all the key components of the problems. Furthermore, each problem has an algebraic solution that corresponds to each diagram and essentially replaces the visual unit with a variable.

The need for proficiency in problems of this type was reinforced by an article in the *Boston Globe* (Vaishnav, 2005) about the "toughest" problem on the 2005 Massachusetts Grade 10 statewide assessment in mathematics. Approximately 50% of the 72,000 test takers answered the following multiple choice question incorrectly.

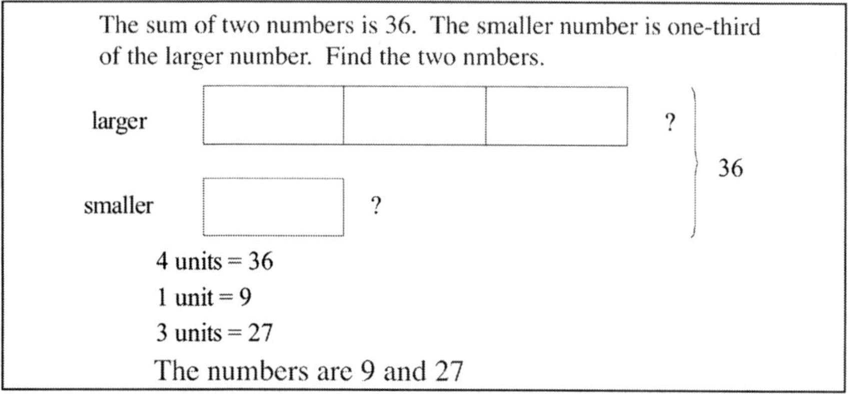

Figure 17.1.   Grade 3 example.

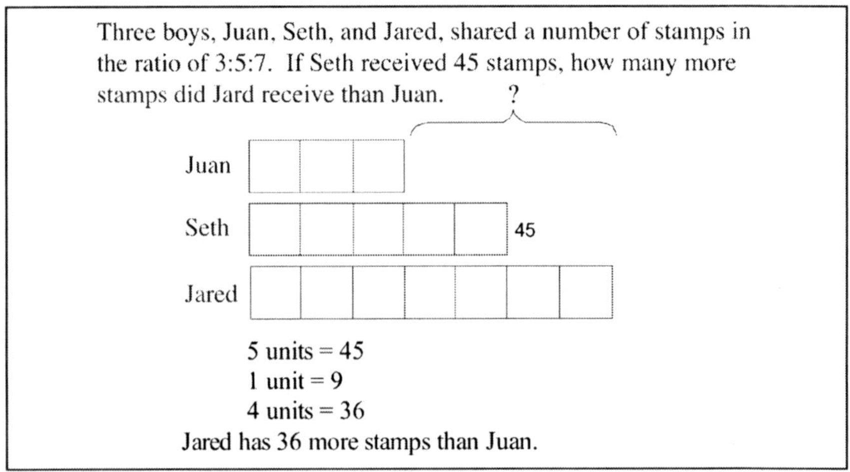

Figure 17.2.   Grade 4 example.

Figure 17.3.   Grade 5/6 example.

Of the people in attendance at a recent baseball game, one third had grand-
stand tickets, one fourth had bleacher tickets, and the remaining 11,250
people in attendance had other tickets. What was the total number of people in attendance at the game?

A) 27,000, B) 20,000, C) 16,000, or D) 18,000.

The problem is typical of many in the Grade 6 Singapore texts. It can
be readily solved using the model drawing technique. When I show the
problem to elementary school teachers, most do not believe they could do
the problem or even understand the solution. Then I show them the solution below and they begin to see the possibility of understanding mathe-

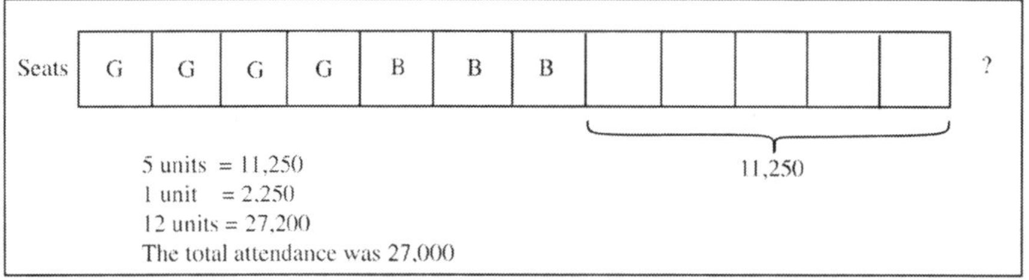

Figure 17.4.   Example of "toughest problem."

matical ideas that have always eluded them. The model drawing solution is shown in Figure 17.4. Note again how the diagram includes the key elements of the question.

MENTAL MATH

Mental math techniques emphasize an understanding of place value and the distributive, commutative and associative properties. They help students develop computational fluency and free them from excessive memorization. On page 72 of the 3A books (Kong, 2003), the rectangular array model for multiplication is used to teach students how to use these properties to help with single-digit multiplication. They are asked think of "6 × 7" as "6 × 5" plus "6 × 2." They can compute "6 × 8" by doubling "6 × 4," and find "6 × 9" by subtracting "60 – 6." Other examples include:

1. $72 - 59 = 73 - 60$

2. $32 \times 25 = (32 \times \frac{1}{4}) \times 100$

3. $35 \times 98 = (35 \times 100) - (35 \times 2)$

4. $142 \div 5 = 284 \div 10$

I will leave it to the reader to consider the mathematics concepts in each example. However, I will note that each can be explained from either a theoretical or applications oriented viewpoint. For instance, the approach shown for 35×98 can be explained either using the distributive property or by way of a story problem where you buy 35 items for $100 each and then discover you have a coupon good for a $2 discount on each item. Students and teachers who are given regular practice with these techniques develop confidence in their abilities because they are based on understanding. Teachers find they now have more ways to explain a concept.

SUMMARY

The American Institute for Research study quotes a teacher who stated: "I never realized that I do not understand math until I had to teach mathematics from the Singapore textbooks" (Ginsburg et al., 2005, p. 1). When teachers see problems in an elementary text that they themselves cannot solve, it might convince them to confront the gaps in their own understanding of mathematics. This alone is not enough. Many of our K–8 teachers are math phobic. By providing strategies for comprehending topics that never made sense to them, these books give teachers a way to become successful mathematics learners. They, in turn, are more likely to teach in a manner that stresses understanding.

REFERENCES

Ginsburg, A., Leinwand, S., Anstrom, T., & Pollock, E. (2005). *What the United States can learn from Singapore's world-class mathematics system (and what Singapore can learn from the United States): An exploratory study.* Washington, DC: American Institutes for Research.

Kong, K. T. (2003). *Primary mathematics 1-6* (U.S. edition). Singapore: Federal Publications.

Meng, S. K., & Yoong, W. K. (1997). *New elementary mathematics syllabus d, Vol. 1-2.* Singapore: Pan Pacific Publications.

National Council of Teachers of Mathematics. (2000). *Principles and standards for school mathematics.* Reston, VA: Author.

Parker, T., & Balderidge, S. (2003). *Elementary mathematics for teachers.* Okemos, MI: Sefton-Ash.

Scott Foresman-Addison Wesley. (2004). *Mathematics.* Upper Saddle River, NJ: Pearson Education.

The University of Chicago School Mathematics Project. (2001). *Everyday mathematics.* Chicago: Wright Group/McGraw Hill.

Vaishnav, A. (2005, August 25). Familiarity may not breed MCAS success. *Boston Globe.* Retrieved June 19, 2008, from http://www.boston.com/news/local/articles/2005/08/25/familiarity may not breed mcas success/

PART IV

STATUS OF CALCULATOR AND
COMPUTER TECHNOLOGY IN THE K–12 CURRICULUM

CHAPTER 18

CALCULATOR AND COMPUTER TECHNOLOGY IN THE K–12 CURRICULUM

Some Observations From a U.S. Perspective

M. Kathleen Heid
The Pennsylvania State University

The nature and implementation of a mathematics curriculum depends on the purpose for which it is intended, and the role of technology in a curriculum is highly dependent on that purpose. Factors influenced by that purpose and the role of technology in carrying out the goals of a curriculum include access to technology, experience with technology, and the opportunities for organization or selection of material afforded by the technology. This paper will note some of the prominent issues that have influenced U.S. implementation of technology in mathematics curricula, drawing on bodies of research and documentation produced by the government, by individual researchers, and by those who have synthesized the literature of the field.

Mathematics Curriculum in Pacific Rim Coutries—China, Japan, Korea, and Singapore: Proceedings of a Conference, pp. 293–303

ACCESS TO TECHNOLOGY

The extent to which technology influences mathematics curricula depends on a range of variables, one of which is students' access to technology. The United States has seen astounding growth over the past decade in school access to technology, epitomized by the rapid growth in public school access to the Internet. Not only are schools wired, but a high percentage of individual classrooms have Internet access and the student to computer ratio is approaching a reasonable ratio. Government reports, summarized by Larry Cuban (2001), noted that in the early 1980s there was an average of 125 students per computer in U.S. schools—a ratio that dropped to 18 to 1 in the 1990s and 5 to 1 by the year 2000. This rapid growth in access also characterized student in-school access to the Internet, with 100% of the nation's public schools and 93% of the nation's public school classrooms having Internet access by 2003 (see Table 18.1; Kleiner & Farris, 2002; Kleiner & Lewis, 2003; Parsad, Jones, & Greene, 2005). By 2003, most (95%) of public schools that had access to the Internet had broadband access, and 48% of public schools allowed students access to the Internet outside of regular school hours.

Consistent with these government reports, Dunham and Hennessy (2008), in their comprehensive analysis of equity issues in technology and the teaching and learning of mathematics, point out that availability of all types of technology has increased dramatically in the last decade and that

Table 18.1.   Internet Access in U.S. Public Schools—A Numerical View

Year	Percent of Public Schools in the U.S. With Access to the Internet	Percent of Public School Instructional Rooms With Internet Access	Ratio of Public School Students to Instructional Computers With Internet Access
1994	35	3	
1995	50	8	
1996	65	14	
1997	78	27	
1998	89	51	12.1
1999	95	64	9.1
2000	98	77	6.6
2001	99	87	5.4
2002	99	92	4.8
2003	100	93	4.4

Source: Data displayed are drawn from Parsad, Jones, and Green (2005), Kleiner and Farris (2002), Kleiner and Lewis (2003).

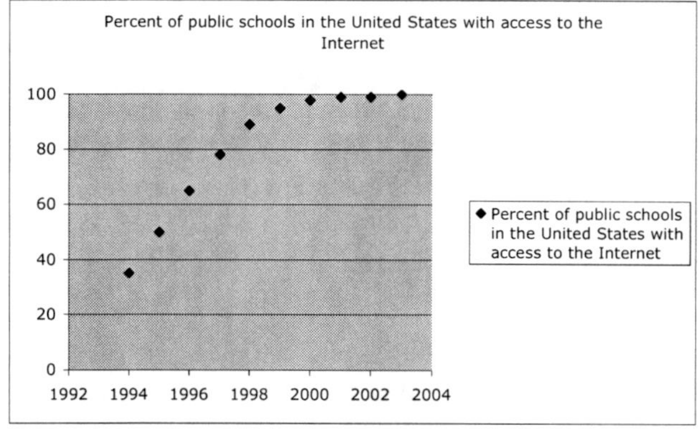

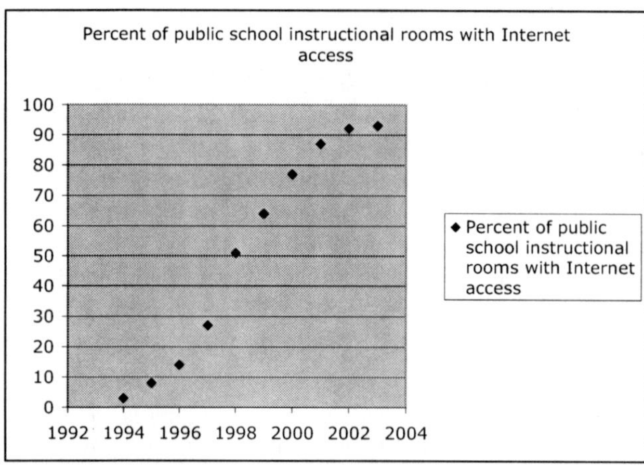

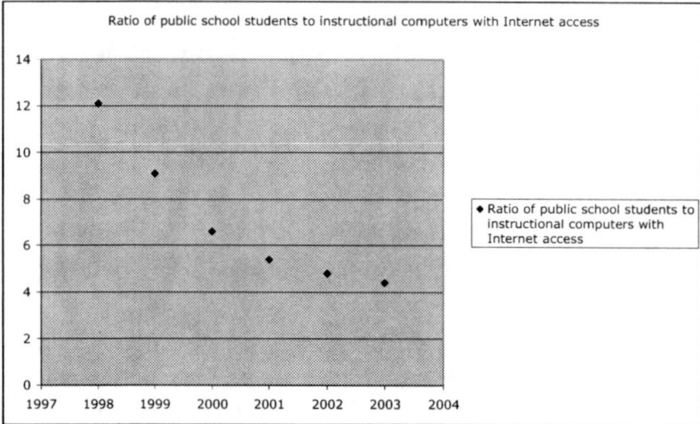

Source: Data displayed are drawn from Parsad et al. (2005), Kleiner and Farris (2002), Kleiner and Lewis (2003).

Figure 18.1.   Internet access in U.S. public schools—a graphical view.

some differences that were once apparent in calculator access across gender, ethnicity, and income have subsided. Dunham and Hennessy observed, however, that "Knowing how many computers or calculators are available in a school does not, of course, tell us how often, in what manner, and by whom the technology is used." It is within differences in these areas that inequities may be visible.

Experience With Technology

Students' experience with technology is not a direct function of the student's potential for access to computers—other factors play a major role in defining the nature of students' technology experiences. Even in schools with excellent access to technology, students may not experience a technology-integrated curriculum. Cuban studied one such situation. Cuban (2001), analyzing technology use across a range of subject areas in a selection of Silicon Valley schools, observed that technology, even if highly available, is often not used to the fullest extent and not well-integrated into the curriculum. Dunham and Hennessy (2008) echo that observation as it applies to use of technology in mathematics instruction. They note that the nature of technology use in the sources for their research synthesis varies among schools and among subjects and that "differential experiences can prevent some groups—minority, lower socioeconomic status, inner-city, rural, disabled, and female students—from benefiting from the full potential of educational technology."

The differential experience across settings is related to teacher factors. Dunham and Hennessy (2008) observed that the ways in which students experience technology is greatly influenced by their teachers' perspectives on and experience with technology. Zbiek and Hollebrands (2008) build a strong research-based analysis of how mathematics teachers' use of technology develops. Beginning with their personal use of technology, teachers' use evolves into their managing technology use in the classroom, and continues to their assessment of student understanding as it emerges in students' use of technology. Some researchers have documented teachers' use of technology in relation to their philosophy of teaching and learning. Akujobi (1995) summarized the results of his dissertation study:

> Analyses of the two clusters showed that teachers who held conceptual views about mathematics taught mathematics in alternative ways and supported the use of technology for instruction. In contrast, teachers who viewed mathematics as a set of rules and procedures tended to avoid using technology and envisioned its use narrowly—for remedial purposes and drill-and-practice.

This observation is compatible with results from Becker's survey (2001) of the ways in which teachers use computers. A 1998 survey of more than 4,000 teachers of students in Grades 4 through 12 suggested that, although mathematics teachers reported greater use of computers than teachers in some other subject areas, frequent use was predominant only for low-ability students in middle schools attended by economically disadvantaged families. Becker concluded that "Computers thus seemed to be primarily used for drilling basic skills to children who are otherwise disadvantaged" (p. 6).

Becker conceptualized two commonly discussed contrasts in types of teaching philosophy and associated teachers with one of these philosophies according to their assessment of a given classroom scenario, The types of teaching philosophy were:

> *Traditional Transmission Instruction* is based on a theory of learning that suggests that students will learn facts, concepts, and understandings by absorbing the content of their teacher's explanations or by reading explanations from a text and answering related questions. Skills (procedural knowledge) are mastered through guided and repetitive practice of each skill in sequence, in a systematic and highly prescribed fashion, and done largely independent of complex applications in which those skills might play some role.
>
> *Constructivist-Compatible Instruction* is based on a theory of learning that suggests that understanding arises only through prolonged engagement of the learner in relating new ideas and explanations to the learner's own prior beliefs. A corollary of that assertion is that the capacity to employ procedural knowledge (skills) comes only from experience in working with concrete problems that provide experience in deciding how and when to call upon each of a diverse set of skills. (Becker, 2001, p. 9)

Becker observed that teachers' philosophies differed considerably by subject matter. Comparing English, social studies, science, and mathematics teachers at both the middle school and high school levels, Becker concluded from his data that high school mathematics teachers (along with middle school social studies teachers) were the most transmission-oriented in philosophy, although there was a range of philosophies within teachers of each subject matter. Comparing teaching philosophies with classroom use of computers, Becker found that teachers with the most constructivist-compatible philosophies "use computers more frequently, use them in more challenging ways, use them more themselves, and have greater technical expertise." He concluded that when "constructivist-oriented" teachers have sufficient resources in their classrooms (five or more computers) as well as a reasonable level of experience and skill in using computers, a majority will have their students make active and regular use of computers during their class period.

It was the frequency of student use that distinguished the most constructivist-compatible teachers. The picture was slightly different for mathematics teachers, however, with moderately traditional teachers using computers the most. Becker's analysis of the mathematics teachers who reported frequent computer use found contrasting objectives held by teachers with these two different teaching philosophies: transmission-oriented teachers had mastering and remediating skills as their major objective (80% and 63%, respectively) for student computer use, while a small percentage of the constructivist teachers held these as their primary goals (15% each). Teachers in Becker's survey with both teaching philosophies had their students using games to practice skills, but the students of constructivist-compatible teachers used a broader range of software for a broader range of goals.

Technology and Learning the Content of School Mathematics

An analysis of the impact of technology on the content of school mathematics is beyond the scope of this paper (see Heid & Blume, 2008, and Blume & Heid, 2008, for extended discussion of research on technology and the teaching and learning of school mathematics.) This paper focuses on only one general characteristic of mathematics-specific technology. A defining characteristic of the most promising mathematics-specific technology used in school settings is their multirepresentational capacity and the fact that those multiple representations are often hot-linked and dynamic. This section will contain a few observations, gleaned from research, about the impact on learning of these multirepresentational environments.

When student are engaging in problem solving, multiple representations can afford them the opportunities to pursue alternative strategies in an effort to circumvent the need to deal with unfamiliar representations (Heid, Hollebrands, & Iseri, 2002; Iseri, 2003). In one study of middle school students learning algebra, Kevin (a mathematically talented seventh grader) exhibited an uncanny ability to use the tools with which he was familiar to interpret unfamiliar representations. Although availability of multiple representations can result in students avoiding work in an unfamiliar representation, work in one representation can also trigger work in (and insight into) another representation of the same mathematical object. Yerushalmy (1997) recounts precalculus students' investigations of asymptotes in a qualitative technological environment; one student's description of a function with a nonhorizontal linear asymptote as "an imaginary line that divides the function in two" (p. 18) resulted in a group discussion of the form of functions rules with similar properties. Yerushalmy observed: "The learner's action on an unexplored object presented in one representation (the graph produced by the software) leads to inquiry about the processes underlying that object explored through a parallel representation (the symbolic procedure)" (p. 22). The immediate access to both symbolic and graphical form supported students' search for a generalization of the symbolic rule.

A curriculum that capitalizes on the multirepresentational features of technology can enable instruction to focus on central mathematical objects and processes. An early example of technology promoting this anchoring focus is the functions-based approach to algebra made possible through integrating the use of (the equivalent of) a computer algebra system into introductory algebra so that students could concentrate their attention on the mathematical object of *function* (Fey et al., 1995/1999). Students using this curriculum developed a more robust understanding of *variable* (Boers-van Oosterum, 1990) and of *function* (O'Callaghan, 1998). Even when compared with students of similar ability who had completed at least two additional years of traditional school mathematics, students engaged in this curriculum developed a greater flexibility in reasoning with multiple embodiments of mathematical functions (Sheets, 1993). Not only can technological multi-representational environments contribute to the development of robust mathematical understanding, but it can also reveal students' less-than-solid understandings of fundamental mathematical concepts such as the distinctions among *function*, *equation*, and *expression*, and the difference between approximate and exact calculations (Guin & Trouche, 1999). The benefits of access to multiple representations is not confined to algebra. Work by Kaput and colleagues (Kaput 1994) demonstrated how technology can be used to plant the mathematical seeds for the ideas of calculus.

TECHNOLOGY AND THE SELECTION AND ORGANIZATION OF THE CONTENT OF SCHOOL MATHEMATICS

While the integration of multirepresentational technology into school mathematics curricula can affect or reveal students' understandings of mathematical objects such as *function* and *variable*, technology can be integrated into mathematics curricula in a more comprehensive way. Such global strategies are particularly interesting when viewed in the context of cognitive tools: tools that "transcend the limitations of the mind" (Pea 1987, p. 91). In mathematics instruction, cognitive tools include tools such as computer algebra systems (CAS), dynamical geometry programs, dynamic statistics programs, microworlds, computer-based laboratories, and graphing utilities, and there is a range of ways that their use in a mathematics curriculum can be characterized. In this section, I will describe some of the ways cognitive tools can substantially affect the nature of mathematics teaching and learning.

One productive way to classify uses of cognitive tools in mathematics instruction is to view them as some combination of amplifiers and reorganizers (Heid, 1997; Pea, 1985). As an amplifier, technology extends the existing curriculum, with little change in the actual content. As a reorganizer, technology can change the fundamental nature and arrangement of the curriculum. For example, computer games that help students practice the skills that are traditionally in the curriculum might be classified as amplifiers, while computer algebra systems that are used to replace or resequence some algebra content might be classified as reorganizers. It is my observation that, by far, the greater use of technology in mathematics curriculum in the United States is as amplifiers, while it well may be that it is through the use of technology as reorganizers that curricula can truly capitalize on the use of technology.

It is the selection, order, emphasis, and depth of development of skills and concepts that lie at the heart of technology's potential effect on mathematics curriculum. Technology can allow the selection of different content in a school mathematics curriculum. The *Concepts in Algebra* curriculum (Fey et al., 1995/1999), for example, was designed to focus beginning algebra students on the concept of function in real world settings. Curricula created through the CAS-Intensive Mathematics Project (Heid, Zbiek, Blume, & Choate, 2004) extends the notion of function as an organizer for secondary mathematics to transformation geometry, geometry of congruence and similarity, iteration, composition and inverse, and mathematical modeling. In both of these curricula, both the content and the approach to the content served to reconceptualize secondary school mathematics.

Technology can be used to orchestrate the grain size of a curriculum's focus on procedures. Computer algebra systems can be used to structure a curriculum in different chunk sizes. Using a CAS, one can focus on equation solving as a whole (a macroprocedure), on selections of the steps involved in equation solving (midiprocedures), or on the execution of equation solving steps (microprocedures). This flexibility in the nature of discussion of procedures can result in a large range of options for a school algebra curriculum—options that might be tailored to the goals of the particular course. The notion that there is more to procedures than their speedy execution is an idea that is gaining prominence in the U.S. discussion of computer algebra systems. Groups like the USA-CAS group have provided forums for the discussion of alternative visions of the role of procedures in school mathematics as well as for the interpretation of the French notions of *instrumentalization* and *technique* in the U.S. setting.

CONCLUSION

This paper highlights progress in getting technology into schools and classrooms and points to areas of needed improvement in the integration of that technology into the experiences of U.S. secondary school students. It discusses research on mathematics teachers' use of technology and on the impact of their educational philosophy on the ways in which they use technology in their teaching. Finally, it points out ways in which particular visions of technology use can affect secondary mathematics curricula. It is through the enactment of visions like those described that U.S. mathematics classrooms have the potential to progress beyond the "all dressed up and no place to go" phenomenon. U.S. mathematics classrooms are to a great extent equipped and ready to go—it is time for mathematics curricula to catch up.

REFERENCES

Akujobi, C. O. (1995). Teachers' knowledge and beliefs about the use of computers in high school mathematics. *Dissertation Abstracts International, 56(07)*, 2594. (UMI No. 9537184)

Becker, H. J. (2001, April). *How are teachers using computers in instruction?* Paper presented at the meeting of the American Educational Research Association, Seattle, WA.

Blume, G., & Heid, M. K. (Eds.). (2008). *Research on technology and the teaching and learning of mathematics: Volume 2. Cases and perspectives.* Charlotte, NC: Information Age.

Boers-van Oosterum, M. A. M. (1990). Understanding of variables and their uses acquired by students in traditional and computer-intensive algebra (University of Maryland College Park). *Dissertation Abstracts International, 51,* 1538A.

Cuban, L. (2001). *Oversold and underused: Computers in the classroom.* Cambridge, MA: Harvard University Press.

Dunham, P., & Hennessy, S. (2008). Equity and use of educational technology in mathematics. In M. K. Heid & G. Blume (Eds.), *Research on technology and the teaching and learning of mathematics: Volume 1. Research syntheses* (pp. 345–418), Charlotte, NC: Information Age.

Fey, J. T., Heid, M. K., Good, R., Blume, G., Sheets, C., & Zbiek, R. M. (1999). *Concepts in algebra: A technological approach.* Chicago: Everyday Learning. (Original work published 1995)

Guin, D., & Trouche, L. (1999). The complex process of converting tools into mathematical instruments: the case of calculators. *International Journal of Computers for Mathematical Learning, 3*, 195–227.

Heid, M. K. (1997). The technological revolution and the reform of school mathematics. *American Journal of Education, 106*(1), 5–61.

Heid, M. K., & Blume, G. (Eds.). (2008). *Research on technology and the teaching and learning of mathematics: Volume 1. Research syntheses.* Charlotte, NC: Information Age.

Heid, M. K., Hollebrands, K., & Iseri, L. (2002). Reasoning, justification, and proof, with examples from technological environments. *Mathematics Teacher, 95*(3), 210–216.

Heid, M. K., Zbiek, R. M., Blume, G. W., & Choate, J. (2004). *Technology-intensive secondary school mathematics curriculum* (Modules 1 to 9). (Available from the authors on CD: Dr. M. Kathleen Heid, The Pennsylvania State University, 271 Chambers Building, University Park, PA 16802).

Iseri, L. (2003). *Algebra students' developing symbolic reasoning in the context of a computer algebra system* (Doctoral dissertation, The Pennsylvania State University). *Dissertation Abstracts International, 64*, 4397.

Kaput, J. (1994). Democratizing access to calculus: New routes using old roots. In A. Schoenfeld (Ed.), *Mathematical thinking and problem solving* (pp. 77–155. Hillsdale, NJ: Erlbaum.

Kleiner, A., & Farris, E. (2002). *Internet access in U.S. public schools and classrooms: 1994-2001* (NCES 2002-018). Washington, DC: U.S. Department of Education, National Center for Education Statistics.

Kleiner, A., & Lewis, L. (2003). *Internet access in U. S. Public Schools and Classrooms: 1994-2002* (NCES 2004-011). Washington, DC: U.S. Department of Education, National Center for Education Statistics.

O'Callaghan, B. R. (1998). Computer-intensive algebra and students' conceptual knowledge of functions. *Journal for Research in Mathematics Education, 29*, 21–40.

Parsad, B., Jones, J., & Greene, B. (2005). *Internet access in U. S. public schools and classrooms: 1994–2003* (NCES 2005-015). U.S. Department of Education. Washington, DC: National Center for Education Statistics.

Pea, R. D. (1985). Beyond amplification: Using the computer to reorganize mental functioning. *Educational Psychologist, 20*, 167–182.

Pea, R. D. (1987). Cognitive technologies for mathematics education. In A. H. Schoenfeld (Ed.), *Cognitive science and mathematics education* (pp. 89–122). Hillsdale, NJ: Erlbaum.

Sheets, C. (1993). *Effects of computer learning and problem-solving tools on the development of secondary school students' understanding of mathematical functions.* Unpublished doctoral dissertation, University of Maryland College Park.

Yerushalmy, M. (1997). Reaching the unreachable: Technology and the semantics of asymptotes. *International Journal of Computers for Mathematical Learning, 2*, 1–25.

Zbiek, R. M., & Hollebrands, K. (2008). A research-informed view of the process of incorporating mathematics technology into classroom practice by inservice and prospective teachers. In M. K. Heid & G. Blume (Eds.), *Research on tech-*

nology and the teaching and learning of mathematics: Vol. 1. Research syntheses (pp. 287–344). Charlotte, NC: Information Age.

CHAPTER 19

THE STATUS OF CALCULATOR TECHNOLOGY IN UNITED STATES K–8 MATHEMATICS CURRICULUM

It Depends On How You Look At It

Kathryn B. Chval
University of Missouri-Columbia

The availability of calculators has influenced mathematics instruction, assessments, and textbooks since calculators were first introduced into K–8 mathematics classrooms 30 years ago. Use of calculators has been supported by a steady line of research (e.g., Hembree & Dessart, 1986; 1992; Shumway et al., 1981; Suydam, 1979) and numerous recommendations from professional organizations (National Council of Teachers of Mathematics, 1989, 2000, 2005; National Research Council, 1990; National Council of Supervisors of Mathematics, 1988). Even though the use of calculators has been encouraged for some time, their use in U.S. elementary- and middle-school mathematics classrooms remains controversial and uneven.

The United States has more than 14,000 largely autonomous school districts (U.S. Census Bureau, 2004), accounting for great variance in

Mathematics Curriculum in Pacific Rim Coutries—China, Japan, Korea, and Singapore: Proceedings of a Conference, pp. 305–316

whether, why, and how often calculators are used from grade level to grade level, and from classroom to classroom. As we consider the status of calculator use in more than 14,000 U.S. school districts, we must be mindful that classroom calculator usage is situated within a complex system which allows autonomy at many levels—district, school and classroom. Ideally, state standards would provide guidance for suitable uses of calculators, mathematics textbooks would include tasks that require thoughtful calculator use, school districts would purchase and maintain the necessary equipment, teachers would be prepared to use calculators effectively, and students would use them appropriately. Unfortunately, this is not the reality in which we work.

When calculators were first introduced in classrooms, many parents and educators were opposed to their use. The following media headlines illustrate the long-standing nature of opposition to classroom calculator use, as well as the depth and persistence of that opposition:

- "Students' use of hand-held calculators in class, even during tests, is proposed," *The Wall Street Journal*, November 14, 1975 (Bishop, 1975);

- "Why Johnny can't add," *Newsweek*, September 24, 1979 (Sewall, 1979);

- "Calculators in schools: Doubts persist," *The Washington Post*, August 15, 1983 (Latimer, 1983);

- "Classroom calculators add to math illiteracy," *The Wall Street Journal*, May 16, 1986 (Saxon, 1986);

- "The calculator crutch," *The New York Times*, September 29, 1991 (Klutch, 1997); and

- "Ditch the calculators," *Newsweek*, November 3, 1997 (Hunsaker, 1997).

Even though resistance to calculator use persisted for many years, an examination of the Lexis-Nexis database of news coverage regarding calculators indicates that in recent years the U.S. media's coverage of calculators has been more positive. The number of articles depicting negative aspects of calculator use has declined. Furthermore, the focus of more recent articles has shifted toward the required use of calculators, how calculators should be used, and the use of calculators on standardized tests, as the following examples illustrate:

- "In 'bold stroke,' Chicago to issue calculators to all 4th-8th graders," *Education Week*, October 14, 1987 (Rothman, 1987, p. 1);

- "Calculators to be required for math classes," *The Patriot Ledger* (Quincy, MA), May 7, 1996 (Hainer, 1996);

- "Calculators become key tool in math education, *"Associated Press*, October 16, 1999 (The Associated Press State & Local Wire, 1999); and

- "Math tools open minds: Aim is to raise exam scores," *Times-Picayune* (New Orleans), February 1, 2005 (Harvey, 2005).

Another encouraging indicator is that 37 states now make references to calculators within state curriculum documents and 40 states allow the use of calculators on some portions of state mathematics assessments.

CALCULATORS IN STATE STANDARDS DOCUMENTS

To understand and describe the extent to which states support use of calculators in elementary and middle school, researchers associated with the Center for the Study of Mathematics Curriculum (CSMC) conducted an analysis of official state mathematics curriculum standards documents including a specific investigation of expected roles of calculators (Chval, Reys, & Teuscher, 2006). Twenty state documents include a discussion of the role of calculators/technology within the introductory material and 32 state documents include the terms calculator and/or technology within a subset of learning expectations. A few of the findings from this analysis are highlighted below.

The most common messages regarding calculators/technology found in the 20 introductory state documents include:

1. Appropriate use of calculators/technology is encouraged.
2. Calculators/technology are commonly used in the workplace and outside of school, therefore students should use these tools to solve problems.
3. Calculators/technology are tools for learning and teaching.
4. Calculators/technology can support increased understanding.
5. The existence of calculators/technology does not diminish the need for computational fluency.
6. Calculators/technology can support effective teaching.
7. Teachers are responsible for appropriate and effective use of calculators/technology.

In summary, each of the 20 state documents which include statements regarding the use of calculators/ technology notes the potential of these tools to support teaching and learning.

Such statements, including grade-by-grade learning expectations, are carefully considered by school administrators and teachers as they design, teach, and monitor mathematics learning. Additionally, such statements are given great weight in the writing and editing of mathematics textbooks. Therefore, these documents and the messages they convey are likely to have substantial impact upon how mathematics is taught, from textbook content to planning and implementation at the district, school, and classroom levels.

In our analysis of state grade level expectations (LEs), we found 31 state documents that included a total of 396 LEs related to the use of calculators among a total of about 14,600 LEs for elementary and middle school (see Reys, Dingman, Sutter, & Teuscher, 2005, for a more complete summary of the documents). In other words, approximately 3% of the mathematics LEs in the 31 state documents examined include a reference to the use of calculators.

For example, the following grade-level expectations illustrate statements related to the use of calculators:

> Identify, model, describe, and evaluate relationships using graphs, with and without technology. (Nevada, Grade 8)

> Develop, with and without appropriate technology, computational fluency in multiplication and division up to two-digit by one-digit numbers using two-digit by one-digit number contextual problems (Alaska, Grade 3)

Table 19.1 summarizes the 396 calculator/technology LEs, sorted by grade for each of the states. It is clear that the number and nature of the expectations vary considerably from state to state. This raises the question as to whether state standards, textbooks, and state assessments are aligned with these different expectations for calculator use.

In summary, 10 of the 42 states represented in Table 19.1 have mathematics curriculum standards documents that contain no references to calculators within the set of grade-level LEs. Another 18 of 42 states include ten or fewer references to calculators within their documents. Only two state documents—those of Arkansas and Washington—include more than 25 LEs that reference calculators across grades K–8. As noted, across all the documents, the largest concentration of references to calculators is in Grades 6–8. In fact, 211 of the 396 (53%) calculator-related LEs identified are found at Grades 6, 7, or 8.

In addition to identifying the number of LEs that reference calculators/technology, the analysis included a review of the intended role of the calculator within the LEs. Table 19.2 shows the different roles.

Further, two other sets of LEs referred to calculators. However, the focus was not on using calculators but rather on judgments made prior to or after use of the tool. They include choosing an appropriate method of calculation and checking the reasonableness of calculated answers. Table 19.3 shows the frequency of these sets of LEs.

The analysis of the state LE data suggests that calculators are infrequently encouraged solely as a computational tool. The most prominent roles for calculators are for developing or demonstrating conceptual understanding (in K–2), for solving problems or equations (in Grades 3–5), and for representing mathematics (in Grades 6–8).

CALCULATORS IN U.S. TEXTBOOKS

Most current mathematics textbooks include mathematical tasks designed for the specific use of calculators. Textbooks either refer to generic use of calculators or include keystroke sequences for specific models. Some textbooks use calculators more appropriately than others. Some use is superficial (e.g., students pull out the calculator for the one word problem that has more complicated numbers) or it is limited to exposure rather than intentional use. A more formal analysis of calculator-related tasks within current U.S. textbooks and investigations into whether and when they are effective during classroom implementation is needed.

IMPLEMENTED CURRICULA

There have been some large-scale studies that have included the frequency of calculator use at various grades (e.g., Knapp, 1995; National Center for Education Statistics, 2000; Weiss, Banilower, McMahon, & Smith, 2001). The frequencies of use vary from study to study, but at the same time indicate that technology use is not widespread in the elementary grades. For example, Malzahn (2002) reported how often K–2 and 3–5 teachers used calculators or computers for various instructional activities and Hudson, McMahon, and Overstreet (2002) reported frequencies for Grades 5-8 (see Table 19.4) using results from the *2000 National Survey of Science & Mathematics Education*.

As shown in Table 19.4, K–2, 3–5, and 5–8 teachers ask students to practice routine computations/ algorithms without the use of calculators more frequently than they require students to use calculators. Frequency of use increases in the upper grades.

From reviewing the literature and visiting classrooms, there are concerns with how calculators are used in large numbers of K–8 classrooms in the U.S. It is not surprising that there are different stages of implementation from non-use, to mechanical use, to refined use, to innovative use. Other concerns include:

Table 19.1. Number of Calculator/Technology Learning Expectations per Grade by State

State	K	Gr. 1	Gr. 2	Total Gr. K-2	Gr. 3	Gr. 4	Gr. 5	Total Gr. 3-5	Gr. 6	Gr. 7	Gr. 8	Total Gr. 6-8	Total Gr. K-8	Mean of State
AL**														
AK					3	2	2	7	3	2	3	8	15	1.67
AR	2	2	2	6	4	5	4	13	6	13	18	37	56	6.22
AZ									1	1	1	3	3	0.33
CA									1			1	1	0.11
CO		1	1	2	1	1	3	5	3	3	3	9	16	1.78
DODEA														
DC*														
FL		3	5	8	2	2	2	6	3	2	3	8	22	2.44
GA		1	1	2	1	2	2	5	2	2	3	7	14	1.56
HI						1	1	2					2	0.22
ID							1	1	1			1	2	0.22
IN									2	1	4	7	7	0.78
KS		1	1	2	2	2	1	5	3	3	4	10	17	1.89
LA					1	1		2			1	1	3	0.33
MD														
ME														
MI									1	1	1	3	3	0.33
MN									3	3	4	10	10	1.11
MO														
MS	1	1		2	3	1	4	8	5			5	15	1.67

NC					1	3	1	5	1	1	1	3	8	0.89
ND					1			1			1	1	2	0.22
NH/RI														
NM			1	1						1	4	5	6	0.67
NJ					4	4	3	11	4	5	5	14	25	2.78
NV	1	1	1	3	1	1	3	5	3	4	7	14	22	2.44
NY							3	3	1	3	1	5	8	0.89
OK												2	2	0.22
OH					1	1	1	3	1	2	2	5	8	0.89
OR												1	1	0.11
SC			3	3	2	2	3	7	1	1	1	3	13	1.44
SD														
TN		1	1	2		1	1	2	2	2	2	6	10	1.11
TX	3	3	3	9	3	2	4	9	1	2	4	7	25	2.78
UT			1	1	1	1	3	5	2			2	8	0.89
VA	1	4	2	7	1	7	5	13	3			3	23	2.56
VT														
WA		2	5	7	4	5	3	12	6	9	7	22	41	4.56
WV										3	5	8	8	0.89
WY														
Total LE	8	20	27	55	36	44	50	130	59	66	86	211	396	396
Mean per grade level	0.26	0.65	0.87	0.59	1.16	1.42	1.61	1.40	1.90	2.13	2.77	2.27	12.77	1.42

*The DC document includes "technology integration" LEs which span all content areas and include emphasis on learning about technology.

**The shaded rows indicate state documents that do not reference calculators or technology within the LE statements.

Table 19.2.   Role of Calculator/Technology as Specified in LEs Within State-Level Documents

Role of Calculator/ Technology	Grade Band	No. of States	No. of LEs	Total LEs*	Percentage of Total LEs (N = 396)
Solve problems or equations	K-2	6	16	130	33
	3-5	15	46		
	6-8	21	68		
Represent	K-2	2	5	105	27
	3-5	11	17		
	6-8	21	83		
Compute or estimate	K-2	2	3	79	20
	3-5	13	31		
	6-8	15	45		
Develop or demonstrate conceptual understanding	K-2	6	19	64	16
	3-5	8	19		
	6-8	11	26		
Describe, explain, justify, or reason	K-2	8	16	63	16
	3-5	8	18		
	6-8	9	29		
Analyze	K-2	2	3	51	13
	3-5	5	7		
	6-8	15	41		

*The number of LEs does not equal to 396 because some LEs were coded in multiple categories.

Table 19.3.   Summary of Learning Expectations Referring to Choosing Appropriate Methods of Calculation and Checking Reasonableness

Tools	Grade Band	No. of States	No. of LEs	Total LEs*	Percentage of Total LEs (N = 396)
Choose appropriate method of calculation	K-2	4	8	78	20
	3-5	15	36		
	6-8	13	34		
Determine the reasonableness of a calculated answer	K-2	2	2	18	5
	3-5	4	9		
	6-8	4	7		

- Technology is not in wide use in the elementary grades.
- When technology is used, it is not necessarily effective.
- Textbooks need additional mathematical tasks that require appropriate use of technology.

**Table 19.4.   Teachers Reported Frequencies of
Various Instructional Activities**

Students Take Part in Instructional Activities	Never/Few Times a Year	1–2 per Month	1–2 per Week	All or Almost All Lessons
Use calculators or computers for learning or practicing skills (K–2)	44%	31%	23%	3%
Use calculators or computers for learning or practicing skills (Grades 3–5)	21%	47%	27%	5%
Use calculators or computers for learning or practicing skills (Grades 5–8)	15%	31%	38%	16%
Use calculators or computers to develop conceptual understanding (K–2)	46%	32%	20%	2%
Use calculators or computers to develop conceptual understanding (Grades 3–5)	32%	43%	21%	3%
Use calculators or computers to develop conceptual understanding (Grades 5–8)	24%	32%	32%	12%
Use calculators or computers as a tool (K–2)	81%	13%	6%	1%
Use calculators or computers as a tool (3–5)	59%	27%	12%	1%
Use calculators or computers as a tool (5–8)	47%	27%	20%	6%
Practice routine computations/algorithms (K–2)	16%	15%	36%	33%
Practice routine computations/algorithms (3–5)	3%	10%	47%	41%
Practice routine computations/algorithms (5–8)	6%	14%	43%	36%

- Technology is constantly changing. Districts have not funded technology at levels adequate to update and maintain equipment.
- Investments in professional development have been insufficient.

Furthermore, resistance continues from the general public, some parents, some teachers, and some administrators. This resistance often includes assumptions that students use calculators regularly in the elementary grades, that calculator use hinders students' mastery of mathematics facts, and that students proficiently computed prior to the introduction of calculators. Evidence suggests these assumptions are misguided.

THE FUTURE: QUESTIONS TO CONSIDER

Thirty years ago the presence of calculators in K–8 classrooms and mathematics textbooks was rare while resistance to calculator use was strong. Over time, the technology became more sophisticated, more cost effective, and more prevalent in K–8 classrooms, while the resistance decreased. Tasks requiring the use of calculators began to appear in K-8 mathematics curricula, state standards, and state assessments. All of these indicators suggest that the status of calculator technology in K–8 mathematics has significantly improved from the time it was first introduced. However, there are also indications, as discussed above that suggest improvement and further investigation are warranted to determine how to best utilize the potential of the calculator in the K-8 mathematics classroom. For example:

- What messages do textbooks implicitly or explicitly communicate about calculators/technology to teachers, students, and parents?
- What are mathematical tasks that make effective and ineffective use of calculators/technology?
- How do we do a better job of preparing teachers to use calculators/ technology during instruction?

A better understanding of the answers to these questions will facilitate more effective uses of calculators in larger numbers of K–8 mathematics classrooms.

REFERENCES

The Associated Press State & Local Wire. (1999, October 16). Calculators become key tool in math education. *The Associated Press State & Local Wire, State and Regional.*

Bishop, J. E. (1975, November 14). Students' use of hand-held calculators in class, even during tests, is proposed. *The Wall Street Journal,* 2 p. 1.

Chval, K., Reys, B., & Teuscher, D. (2006). What is the focus and emphasis on calculators in state-level K-8 mathematics curriculum standards documents? *Mathematics Education Leadership Journal, 9*(1), 3–13.

Hainer, P. (1996, May 7). Calculators to be required for math classes. *The Patriot Ledger,* p. 15.

Harvey, C. (2005, February 1). Math tools open minds: Aim is to raise exam scores. *Times-Picayune* (New Orleans), p. 1.

Hembree, R., & Dessart, D. J. (1986). Effects of hand-held calculators in precollege mathematics education: A meta-analysis. *Journal for Research in Mathematics Education, 17,* 83–99.

Hembree, R., & Dessart, D. J. (1992). Research on calculators in mathematics education. In J.T. Fey (Ed.), *Calculators in mathematics education: 1992 yearbook of the National Council of Teachers of Mathematics* (pp. 22–31). Reston, VA: National Council of Teachers of Mathematics.

Hudson, S., K. McMahon, and C. Overstreet. 2002. *The 2000 National Survey of Science and Mathematics Education: Compendium of tables*. Chapel Hill, NC: Horizon Research.

Hunsaker, D. (1997). Ditch the calculators. *Newsweek, 130*(18), 20.

Klutch, R. J. (1991, September 29). The calculator crutch. *The New York Times*. Retrieved June 20, 2008, from http://query.nytimes.com/gst/fullpage .html?res=9D0CE2DB133AF93AA1575AC0A967958260&scp=1&sq =%22calculator+crutch%22&st=nyt

Knapp, M. and Associates. 1995. *Teaching for meaning in high-poverty classrooms*. New York: Teachers College Press.

Latimer, L. Y. (1983, August 15). Calculators in schools: Doubts persist. *The Washington Post*, p. B1.

Malzahn, K. A. (2002). *The status of elementary mathematics teaching*. Chapel Hill, NC: Horizon Research, Inc.

National Center for Education Statistics. (2000). *National Assessment of Educational Progress, 2000, mathematics assessment*. Retrieved June 20, 2008, from http:// nces.ed.gov/nationsreportcard/

National Council of Supervisors of Mathematics. (1988). *Essential mathematics for the 21st century: The position of the National Council of Supervisors of Mathematics*. Lakewood, CO: Author.

National Council of Teachers of Mathematics. (1989). *Curriculum and evaluation standards for school mathematics*. Reston, VA: Author.

National Council of Teachers of Mathematics. (2000). *Principles and standards for school mathematics*. Reston, VA: Author.

National Council of Teachers of Mathematics. (2005). *Computation, calculators, and common sense: A position statement of the National Council of Teachers of Mathematics*. Reston, VA: National Council of Teachers of Mathematics.

National Research Council. (1990). *Reshaping school mathematics: A philosophy and framework for curriculum*. Washington, DC: National Academy Press.

Reys, B., Dingman, S., Sutter, A., & Teuscher, D. (2005). *Development of state-level mathematics curriculum documents: Report of a survey*. Retrieved June 20, 2008, from http://www.mathcurriculumcenter.org/resources/ASSMReport.pdf

Rothman, R. (1987). In 'bold stroke,' Chicago to issue calculators to all 4th-8th graders. *Education Week, 6*, 1.

Saxon, J. (1986, May 16). Classroom calculators add to math illiteracy. *The Wall Street Journal*, p. 24.

Sewall, G. (1979). Why Johnny can't add. *Newsweek, 94*(13), 72.

Shumway, R. J., White, A. L., Wheatley, G. H., Reys, R. E., Coburn, T. G., & Schoen, H. L. (1981). Initial effect of calculators in elementary school mathematics. *Journal for Research in Mathematics Education, 12*, 119–141.

Suydam, M. N. (1979). *The use of calculators in precollege education: A state-of-the-art review*. Columbus, OH: Calculator Information Center.

U.S. Census Bureau. (2004). Retrieved June 20, 2008, from http://ask.census.gov/

Weiss, I. R., Banilower, E. R., McMahon, K. C., & Smith, P. S. (2001). *Report of the 2000 national survey of science and mathematics education*. Chapel Hill, NC: Horizon Research.

PART V

ROLE OF TESTING IN THE K–12 CURRICULUM

CHAPTER 20

MATHEMATICS ASSESSMENTS

Do They Tell Us the Same Thing

William H. Schmidt
Michigan State University

One of the characteristics that distinguishes U.S. educational practice from that in other countries is the proliferation of test taking at all levels of the system. With the advent of No Child Left Behind states have or are in the process of developing state assessments in mathematics at grade levels three through eight. Additionally many states have assessments at the high school level, some of which have become exit exams necessary for graduation certification. Standing independent of these is the National Assessment of Educational Progress (NAEP). In addition, districts often administer standardized tests that are commercially produced.

Do all of the various mathematics assessments at a given grade level, say eighth grade, give us the same information about what eighth graders know about mathematics—the whole point of such assessments? At one level the answer is obvious. They do not. Recent studies have shown that the estimated percent of students at the proficiency level within many states based on their state mathematics assessments is not the same as the estimated percent derived from the NAEP test (Musick, 2000). To many

Mathematics Curriculum in Pacific Rim Coutries—China, Japan, Korea, and Singapore: Proceedings of a Conference, pp. 319–326

the interpretation of this is that NAEP is a more difficult test than many state assessments or that the criterion for achieving proficiency on some state tests is lower than that of NAEP. What is not addressed by these explanations is the substantive meaning of "more difficult" or "lower criterion." So the central question remains—do these different tests give us the same information about what students know. The public and very often the policy community's perceptions are that they do. This paper addresses the question from the substantive point of view by looking at the implied content domains undergirding the different assessments.

In the United States, at least, many educational researchers, psychometricians, and policymakers believe in or at least act in ways consistent with what can be called the content homogeneity assumption. They believe standardized tests of mathematics at the same grade level have roughly similar content and can be thought of as providing similar information about student achievement as long as the traditional technical requirements are the same. In fact, the technical requirements usually associated with standardized testing procedures are often viewed as assuring such comparability.

The argument presented here is that at the level of content definition, different mathematics tests cannot be assumed to provide comparable information about what students know about mathematics because different tests very often contain different mathematics content. The assumption that the content on a fourth grade assessment is the same as another fourth grade assessment is not borne out in reality. It does matter who the test publisher is or in which state the assessment was taken. To assume otherwise is to imperil substantive interpretations of what the results might mean with respect to reform efforts or to the evaluation of different instructional approaches or materials.

One of the great abuses resulting from incorrectly making the content homogeneity assumption is that some states doing value-added research vertically scale different tests across grades believing that they now have a common scale of mathematics achievement which has substantive implications of what mathematics students know and do not know. The results are certainly questionable if not fundamentally misleading because the content domains of the different assessments are not the same and so the substantive meaning of the new scale is unknown (Martineau, 2006).

Mathematics tests do not in general provide the same information because they have different content profiles associated with them. School mathematics comprises different subdomains. Some of the more obvious ones related to K–12 schooling include: whole number arithmetic, fractions and decimals, basic number theory, algebraic equations, functions, 2D coordinate geometry, 3D geometry, analytical geometry, trigonometry, probability, statistics, and calculus. The particular configuration of content making up an assessment can be drawn in different ways from the subdomains. They can vary as to which of these subdomains are included or excluded as well as to the relative percent of emphasis given to the various subdomains that are included. Not only do they vary at this level, but even within a subdomain. For example, the fractions subdomain includes topics such as equivalent fractions, conceptual models of fractions, common denominators, and basic operations, each of which is based on related but different kinds of knowledge.

In addition to the above, which describes different topics, the expectation of what the student is to do with the content (what we have termed performance expectations) can also vary. Students could be expected to just memorize some aspect of mathematics such as the multiplication tables, use simple algorithms such as how to multiply a three-digit number by a two-digit number, or engage in mathematical reasoning such as doing formal proofs in geometry. Both of these dimensions are part of the content specifications associated with a mathematics assessment.

In short, different mathematics assessments do not typically provide the same information and as a result should not be thought of as providing interchangeable information about student achievement. They vary both in their topic coverage and in their level of performance expectation associated with each of the topics.

SOME EXAMPLES

Several examples are presented to illustrate just how different mathematics assessments can be from the content point of view and as a result why they do not, in general, provide the same information. The document analysis methodology used to generate the data presented in the following examples was developed as a part of The Third International Math & Science Study (TIMSS) (Schmidt, McKnight, Valverde, Houang, & Wiley, 1997). The methods are based on a framework developed internationally which has the two dimensions described previously—topics and performance expectations. We focus in these examples only on the topic dimension of the content framework (Schmidt, McKnight, Houang, Wang, Wiley, & Cogan, 2001).

Table 20.1 shows the results for two different test publishers with regard to their assessments for each of Grades 2 through 9. Even in the earliest grades where some might question the argument being advanced in the paper—that of the lack of content homogeneity across tests—there are differences in the content profiles of the two publishers' tests. Publisher A's third-grade test is made up of items, three fourths of which focus on whole number operations while the test at the same grade level

Table 20.1.   Content Coverage of Tests by Two Publishers

| | Grade | | | | | | | | | | | | | | | |
| | 2 | | 3 | | 4 | | 5 | | 6 | | 7 | | 8 | | 9 | |
Topic	A	B	A	B	A	B	A	B	A	B	A	B	A	B	A	B
Whole number meaning	14%	16%	10%	10%	12%	8%	7%	4%	4%	6%	5%	0%	3%	1%	2%	0%
Whole number operations	65%	49%	75%	42%	68%	42%	74%	16%	39%	20%	22%	11%	29%	8%	28%	2%
Common fractions	1%	4%	2%	1%	8%	3%	15%	16%	21%	19%	21%	20%	21%	17%	24%	0%
Decimal fractions	0%	0%	4%	4%	12%	6%	12%	14%	19%	10%	21%	11%	19%	10%	17%	2%
Measurement units	7%	12%	11%	14%	24%	10%	14%	10%	5%	9%	5%	11%	5%	7%	4%	10%
Equations and formulas	14%	9%	0%	22%	0%	12%	0%	29%	7%	26%	6%	32%	6%	19%	5%	37%
Data representation & analysis	11%	7%	11%	6%	23%	7%	23%	13%	9%	11%	15%	15%	12%	10%	15%	25%

from publisher B has only about 40% such items. The latter test has 22% of its items focused on prealgebra while there are virtually no such items in test A. (The column percents in Table 20.1 do not add to 100 because only a select set of topics were included in the table.)

Consider the following thought experiment. Imagine it is possible a priori to equate the difficulty of the items given the topic and the performance expectations. This would mean that for a given topic, the difficulty of the items across the two tests would be equal. This could be called the conditional difficulty of the items given the topic. This could be done across all topics found in the two tests.

If we can achieve this conditional equality, that the two tests are equal in difficulty given topic coverage, then the difference in difficulty between test A and test B is now only due to the inherent difficulty associated with the mathematics topics themselves and their relative amount of coverage.

The point is, even under these ideal conditions (equal difficulty conditional on topic and performance expectation), where conditional on topic the difficulty of the two tests is equal, the information derived from these two tests are still not the same because of the difference in content profiles.

We argue that the concept of difficulty as typically used should not be confused with information obtained. Many would agree on the face of it, that the answer to the question posed in this paper is no—different assessments are not the same. The basis on which their agreement would likely rest is that the tests are different in difficulty. That does not answer the question, however, as to whether the information provided by the two tests is the same. As we have seen even when difficulty is assured conditional on topic the two tests could still provide different information on student knowledge given the particular choice of topics and performance expectations on the two tests. This implies that arguing that tests provide different information because they have different difficulties is flawed since the differences in difficulties could be due to the lack of conditional equality or to the inherent differences in the particular mathematics topics included or both.

The pattern of differences between test publishers A and B in content profiles exists across the other grades as well. In general publisher A focuses largely on whole number operations and fractions at the later grades. By contrast, the test of publisher B is more about prealgebra, algebra, and fractions. By ninth grade almost 40 percent of the test B focuses on algebra, 25% on data, and only 2% on whole number arithmetic, fractions and decimals. By contrast test A is composed of 5% algebra, 15% data and 70% whole number arithmetic, fractions, and decimals.

The above tests were standardized tests put out by two major publishers. Next we examine state assessments at fourth and eighth grade across six states. Large cross state differences are evident at both fourth and eighth grade. In this analysis we examine only a select set of topics focusing on geometry, algebra, and data.

Consider fourth grade and the topic of data. Coverage of this topic ranges from less than 10% to as much as a third of the test. The same large difference exists across states on the topic of 2D geometry involving polygons and circles ranging from as little as four percent to almost 20%.

One state assessment devoted half of its fourth-grade test to data and probability and another 24% to geometry. A different state assessment had only about 10 percent of its items on each of geometry and data. This contrast is particularly illustrative of the questionable nature of the content homogeneity assumption across assessments as both of these topics are quite distinct from each other and also from the rest of typical fourth grade mathematics. Results from these two state assessments would carry quite different bits of information.

Eighth-grade differences are just as striking. This is the grade where algebra or at least prealgebra enters the curriculum in a serious way. One state devotes almost a third of its state assessment to equations while another only one item. Big differences also exist across these six states in the coverage of 3D geometry, proportionality, relations and functions, and data.

CONCLUSIONS

This paper addresses the use of mathematics assessments to gauge the achievement of students. There are many such assessments both in terms of commercialized standardized tests and state assessments. The American public often thinks of these as interchangeable tests the results of which pretty much give us the same information.

We have argued that this is not the case, that such assessments do not yield the same information about students' understanding of mathematics since the tests do not share a common content profile. Such differences in content are not trivial and should not be ignored in the interpretation of the assessment. Because of this they in general should not be used for program assessment or for deciding curriculum reform unless the content profiles of the assessment match that of the program or curriculum standards.

Describing tests as more or less difficult although useful in some contexts is not the way these assessments should be viewed, but rather as to what types of information each assessment provides. The question should not only be what percent of the students are proficient, but additionally, proficient with respect to what aspects of mathematics, the latter being

Table 20.2.   Content Coverage of Fourth- and Eighth-Grade Assessments From Six States

Topic	State A		State B		State C		State D		State E		State F	
	4	8	4	8	4	8	4	8	4	8	4	8
Geometry: Position, Visualization, & Shape												
• 2-D coordinate geometry			6%	5%		3%	2%	3%	3%	3%	1%	2%
• 2-D geometry: basics	3%	5%	15%	5%		1%		2%		5%	1%	2%
• 2-D geometry: polygons & circles	11%	3%	19%	18%	4%	4%	11%	15%	15%	14%	4%	6%
• 3-D geometry	2%		2%		2%	2%	13%	3%	5%	14%		1%
Geometry: Symmetry, Congruence, & Similarity												
• Transformations	3%			5%	1%	1%	3%	3%	3%		3%	1%
• Congruence & similarity	1%	1%			1%	3%	1%	1%	4%		1%	
• Constructions with straightedge & compass	4%								1%			
Proportionality												
• Problems	0%	11%	6%	18%		4%		8%		3%		8%
• Functions, relations, & equations												
• Patterns, relations, & functions	8%	22%	17%	10%	28%	22%	3%		12%	8%	4%	2%
• Equations & formulas	3%	1%	11%	15%	1%	5%	8%	14%	3%	19%	24%	31%
Data Representation, Probability, & Statistics												
• Data representation & analysis	34%	47%	26%	30%	9%	8%	12%	12%	19%	14%	9%	12%
• Uncertainty & probability	16%	16%	9%	5%	1%	2%	3%	7%	4%	8%	3%	5%

much more useful information and only accessible by examining the content profile of the assessment.

REFERENCES

Martineau, J. A. (2006, Spring). Distorting value added: The use of longitudinal, vertically scaled student achievement data for growth-based, value-added accountability. *Journal of Educational and Behavioral Statistics,* 35-62.

Musick, M. D. (2000). *Setting educational standards high enough.* Retrieved June 20, 2008, from www.sreb.org/main/highschools/accountability/ settingstandardshigh.asp

Schmidt, W. H., McKnight, C. C., Houang, R. T., Wang, H. C., Wiley, D. E., & Cogan, L. S. (2001). *Why schools matter: A cross-national comparison of curriculum and learning.* San Francisco: Jossey-Bass.

Schmidt, W. H., McKnight, C. C., Valverde, G. A., Houang, R. T., & Wiley, D. E. (1997). *Many visions, many aims: A cross-national investigation of curricular intentions in school mathematics.* Dordrecht, The Netherlands: Kluwer Academic Press.

CHAPTER 21

SOME IMPACTS OF TESTING ON MATHEMATICS CURRICULUM FROM K–12

Perspectives

Chris Cox
Kalamazoo Public Schools

It has often been said that many students suffer from test anxiety. Increasingly, students' teachers and principals are feeling anxious as well about tests. School staffs, now more than ever, are held increasingly accountable as student achievement is measured by federally mandated state developed assessments required in Grades 3–8 under the No Child Left Behind (NCLB) legislation (NCLB, 2002). Under this legislation, school stakeholders at all levels have an increased responsibility to ensure that each and every student, school and district is making adequate yearly progress (AYP). What follows is a look at some of the impacts of testing on mathematics curriculum in the state of Michigan.

In 2005, a school's AYP in the state of Michigan was determined by a variety of indicators, such as reaching a target percentage of proficient

Mathematics Curriculum in Pacific Rim Coutries—China, Japan, Korea, and Singapore:
Proceedings of a Conference, pp. 327–334

students, participation in state testing, and school attendance. These indicators apply to the school as a whole and also to qualifying subgroups of students. Qualifying subgroups of students exist when a school has 30 or more students within any of the following eight major classes: American Indian/Alaska native, Asian or Pacific Islander, Black, White, Hispanic, multiracial, limited English proficient, students with disabilities or economically disadvantaged. In order for a school to achieve AYP, the school must show that each and every qualifying subgroup has achieved yearly targets in both mathematics and English language arts. As a result, those schools that reflect the diversity that the United States cherishes are the ones that have the greatest challenge to ensure that all of their subgroups are making adequate yearly progress. In fact, the districts and buildings with the most diverse student populations have the greatest likelihood to not achieve AYP due to the quantity of indicators they must satisfy.

Each state has the autonomy to determine the process in which AYP is calculated and determined. Consider Table 21.1, which illustrates 50 criteria that the Michigan Department of Education (MDE) considers when determining whether a school has achieved AYP. If any one of the indicators are not met, the school does not make AYP.

Schools that do not make AYP are given a set of sanctions starting in the second year of not making AYP. Each year thereafter increases the severity and complexity of sanction, ultimately resulting in the replacement of school staff and closure of the school. Table 21.2 highlights the sanctions a school faces for not making AYP for four years in the State of Michigan. The third option, "Significantly decrease management authority at school," has appeared on the radar of many school principals. Even though a principal may be in the process of substantial and long lasting school improvement, he or she is faced with the short timelines that are demanded in the current implementation of the NCLB legislation. As a result, principals are faced with balancing long-term effective school reform with creating immediate increases in the achievement of their students.

Given the dire need to perform well on the state assessments, many districts and schools are faced with decisions regarding their mathematics program. Such decisions are based on questions such as: (1) How do the state mathematics strands, standards and benchmarks align with a district's curriculum and the district's philosophy of teaching and learning mathematics; (2) Which instructional materials will best support the work of the district as it attempts to achieve the expectations of the state; (3) How can a district best evaluate their efforts in meeting state expectations prior to the actual state assessment; and (4) How does a district balance time for instruction and time for assessment when time is such a

Table 21.1. Michigan Department of Education Sample of AYP Criteria

Group and Subgroups of Students	Met Yearly Proficiency Target*		Had at Least 95% of Students Participate in Assessment		Attendance or Graduation Measure
	Mathematics	English Language Arts	Mathematics	English Language Arts	
Whole school					
American Indian/ Alaska native					
Asian/Pacific Islander					
African American/Black					
White					
Hispanic					
Multiracial					
Limited English proficient					
Students with disabilities					
Economically disadvantaged					

Source: Michigan Department of Education (2003).
Note: *For 2005, the mathematics proficiency target for elementary schools was 56%; middle schools was 43%; high schools was 44%.

Table 21.2.   MDE Sanctions for Schools Not Making AYP for 4 Years

A Title I school that does not make AYP for four consecutive years is identified for corrective action. The school district must continue to offer the transfer option and supplemental educational services. The district must also take at least one of the following actions to improve student academic achievement in the school:

- Replace the school staff who are relevant to the failure to make AYP.
- Implement a new research-based curriculum and provide appropriate professional development for all relevant staff.
- Significantly decrease management authority at the school.
- Appoint an outside expert to advise the school on revising its school improvement plan to address the issues underlying its continued achievement problems.
- Extend the school year or the school day.
- Restructure the internal organization of the school

Source: Michigan Department of Education (2005b).

scarce commodity? There are many curricular implications as districts ponder the answers to these questions.

ALIGNMENT AND MATERIALS

The choices surrounding curriculum and instructional materials are often guided by what state level tests are assessing. Most tests are designed around an assessment content framework. In the State of Michigan, the recently unveiled *K–8 Grade Level Content Expectations for Mathematics* (GLCEM) (MDE, 2005a) provides such an assessment content framework for elementary and middle school mathematics. The GLCEM defines specific student expectations for mathematics that are to be assessed at each grade level in Grades K–8. A similar document is currently being drafted for high school mathematics expectations. The expectations are small grain-sized curricular statements that have inspired assessments that are oriented more towards assessing isolated skills as opposed to assessing complex mathematical processes. Current testing formats fit within the confines of multiple-choice questions. By design, they are not formative assessments or intended to offer diagnostic student information.

This assessment of skill seems to perpetuate traditional rote methods of teaching that easily inhibit innovative instructional practice. While the GLCEM "represent a challenge toward which to aspire" (MDE, 2005a), many teachers have found that the challenge has been to simply shift traditional isolated skills to earlier grades. When mathematical skills, concepts and ideas are assessed in isolation, they will be taught in isolation. Many districts that have developed a district mathematics program based on the content and process standards of the *Principles and Standards for School Mathematics* (PSSM) (National Council of Teachers of Mathematics, 2000) are finding themselves in a quandary as they attempt to ensure that the state grade-level assessable content becomes part of the district's grade level instructional program. The PSSM framework offered mathematics standards within grade bands; whereas, the GLCEM present expectations at each grade. At this moment, very few, if any, sets of research-based published materials completely align with the GLCEM. Consequently, many districts are considering ways in which to reconfigure or resequence textbooks and other instructional materials to align with the assessment trajectory of the GLCEM.

Determining what mathematics content should be tested and when is a challenge that each state and district must confront. Under the grade-level testing mandate of the NCLB legislation, states must have their grade-level expectations reviewed by an external auditor to validate the quality of their standards. Like Michigan, many states have selected

Achieve as their external reviewer. Achieve is an organization developed as an outgrowth of the National Governor's Association. Their presence in working with state education officials has grown greatly in the last few years. While this organization greatly influences state level policies, it is surprising that they made little reference to the NCTM or the PSSM in its review of the GLCEM, *Review of Michigan's Grade-Level Content Expectations* (Achieve, 2003).

With the presence of Achieve, districts in Michigan are faced with three competing "curriculum authorities" that they now must attend to as they develop their district mathematics program. First and foremost would be the curriculum as defined in the textbooks and instructional materials that are a part of the district's mathematics program. Many of the recently published curriculum materials have been developed based on a substantial research base as well as set of principles and learning theories about how students learn mathematics. Second, the NCTM authored curriculum and professional standards (NCTM, 2000) have provided a vivid description of what high quality mathematics teaching and learning should look like. Third, the growing influence of Achieve in state mathematics assessment and curriculum development is evident with the release of their own set of *K–8 Mathematics Expectations* (Achieve, 2004) and its influence on the state expectations it reviews. Each of these entities seeks to develop a coherent mathematics experience for students that differ from one another. These differences are left in the hands of district leaders who must sort, value and agree upon as they create their district mathematics program.

ONGOING EVALUATION

State testing is not comprehensive. State testing tends to be more summative or evaluative rather than formative or diagnostic. Under the existing testing framework in Michigan, no opportunity exists to formally reassess students on specific grade-level expectations after the initial grade-level test.

In Michigan and in many other states, it is important to view the mathematics content of state assessments as a subset of a coherent mathematics curriculum. The existing focus of assessing students mathematical understandings based on specific expectation statements (as opposed to strands, standards or benchmarks) limits the scope by which one looks at processes of learning mathematics such as problem solving, reasoning and proof, communication, connections and representation. These process standards have become equally valuable in the teaching and learning of mathematics as the content standards themselves since the release of the PSSM (NCTM, 2000).

Consequently, many districts construct periodic formative assessments that can help measure the effectiveness of curricular efforts while sampling a student's mathematical abilities throughout a year prior to the fall state assessment. This can be quite a daunting task for a district to tackle. A well-defined district mathematics program ought to also include aligned formative assessments to monitor students mathematical abilities so that come the time of the state assessment there are few if any surprises. However, given the time allocated for mandated testing, districts find it difficult to weave such formative assessments into limited instructional time.

TIME

While testing can help inform instruction and programs, it can also become distracting to instruction. In the state of Michigan, students in Grades 3–8 take four separate tests for mathematics and a similar number for English language arts each year. Since schools are seeking optimal performance from their students, many schools offer a single test in a one-hour block of time with only one sitting per day. As an example, on those days when mathematics is assessed, the teaching of mathematics is forgone. Over the 7 years of mandated testing in Grades 3–8, a student might potentially lose 28 days of mathematics instruction.

In addition to the actual time involved with the state assessments, one must consider the time devoted to preparing for the test. While many districts discourage shifting meaningful instructional time to state test preparation, many teachers feel personally responsible for their students' performance on the test and feel obligated to allocate some classroom time to test preparation. In a case study of a California school, Jo Boaler found that teachers feel the "need to spend more time on test-taking skills, even though they do not believe that this will improve the students' understanding of mathematics" (Boaler, 2003).

In a recent Phi Delta Kappa/Gallup Poll of the public's attitudes toward the public schools, 58% of the respondents felt that the current emphasis on standardized tests encourage teachers to "teach to the tests," that is to say, concentrate on teaching their students to pass the tests rather than teaching the subject (Rose, 2005). While testing has become a mainstay in public education and its relationship to good instruction is integral, states need to further consider less intrusive ways to evaluate the achievement of students.

CONCLUSION

There are many facets of testing that impact mathematics curriculum. Whether it is ensuring state assessable content is a part of a district's mathematics program, considering the cognitive demands and mathematical processes being assessed, or the frequency and length of assessments testing impacts the intended and enacted curriculums. The evolution of state grade level testing as it is today has allowed curriculum to become more transparent. More educators are engaged in meaningful conversations about the teaching and learning of mathematics. The specificity of the grade level expectations in mathematics has moved many teachers who have become complacent over the years.

On the other hand, state testing systems as they are today have opportunities to improve. As a district, as a state, and as a nation, we must continue to consider the relationship between assessment and instruction while striking a balance in a way that improves the teaching and learning of mathematics for all teachers and all students. We must continue to analyze the mathematics content, processes, and contexts that are used in assessing students' mathematical understandings in order to find more agreement on acceptable measures to gauge the mathematics achievement growth of our students while informing our own instructional growth.

REFERENCES

Achieve. (2003). *Review of Michigan's grade-level content expectations.* Retrieved June 20, 2008 from http://www.achieve.org/files/MI-FullReport10-015-04.pdf

Achieve. (2004). *Mathematics achievement partnership; Draft k-8 mathematics expectations.* Retrieved June 20, 2008, from http://www.achieve.org/dstore.nsf/Lookup/MAPK-8fullreport/$file/MAPK-8fullreport.pdf

Boaler, J. (2003). A special section on high-stakes testing: When learning no longer matters: Standardized testing and the creation of inequality. *Phi Delta Kappan, 84*(7), 502–506.

Michigan Department of Education. (2003). *Consolidated state application accountability workbook.* Retrieved June 20, 2008, from http://www.michigan.gov/documents/NCLB_Michigan_Accountability_Workbook_Revised_63716_7.pdf

Michigan Department of Education. (2005a). *K-8 grade level content expectations for mathematics.* Retrieved June 20, 2008, from http://www.michigan.gov/glce

Michigan Department of Education. (2005b). *Requirements for schools not making AYP.* Retrieved June 20, 2008, from http://www.michigan.gov/mde/0,1607,7-140-22709_22875-85932--,00.html

National Council of Teachers of Mathematics. (2000). *Principles and standards for school mathematics.* Reston, VA: Author.

No Child Left Behind Act, 115 Stat. 1428. 2002. Retrieved June 20, 2008, from http://www.ed.gov/nclb/accountability/schools/accountability.html#5
Rose, L. C., & Gallup, A. M. (2005). The 37th annual Phi Delta Kappa/Gallup poll of the public's attitudes toward the public schools. *Phi Delta Kappan, 87*(1), 41–57.

PART VI

REFLECTIONS BY CONFERENCE ATTENDEES

CHAPTER 22

MOVING BEYOND MYTHS TO FOSTER INTERNATIONAL COLLABORATION

International Conference a Step in the Right Direction

Diane L. Moore
Western Michigan University

Jill Newton
Michigan State University

Dawn Teuscher
University of Missouri-Columbia

The federal No Child Left Behind Act of 2001 (Public Law 107-110), requiring schools to help all students be proficient in core areas such as mathematics, has both heightened the concern about student learning and fueled the debates over curricula materials and teaching strategies that support student learning. International studies such as the Program

Mathematics Curriculum in Pacific Rim Coutries—China, Japan, Korea, and Singapore: Proceedings of a Conference, pp. 337–349
Copyright © 2008 by Information Age Publishing

for International Student Assessment (PISA), the Third International Mathematics and Science Study (TIMSS), and TIMSS Repeat (TIMSS-R) have compared student achievement in the United States with that of other countries. Results consistently rank achievement in mathematics for students in the United States far below that of students in Asian countries, such as Singapore and Japan (Mullis, Martin, Gonzalez, & Chrostowski, 2004; Watanabe & McGaw, 2004). The question of how curriculum impacts student performance as reported on these international assessments is certainly worth asking in a conference such as this.

> By focusing on countries with national curricula from one area of the world, the center [CSMC] hopes to clarify the various ways in which countries with similar traditions operate to determine their courses of study and how they translate those courses of study into materials for teachers and students. (Harms, 2005)

This paper offers a reflective glance at what was shared at the conference, focusing on the examination of three myths regarding the teaching and learning of mathematics in China, Japan, Korea, and Singapore compared with the teaching and learning of mathematics in the United States.

A myth is "a belief whose truth or reality is accepted uncritically" (Singham, 2003, p. 586). As the results of international assessments are released, myths emerged from interpretations of the reports. Several such myths were discussed during the conference; however, we will focus on three: (1) students in Pacific Rim countries value mathematics more than students in the United States; (2) teachers in Pacific Rim countries use "traditional" teaching strategies while U.S. teachers use "reform" strategies; and (3) students in Pacific Rim countries rarely use calculators when learning mathematics while American students overuse calculators for computation. Myths in general, are difficult to understand and break down because they are not necessarily completely false. In this paper we outline what is happening in the four Pacific Rim countries and in the United States and discuss why these particular myths need to be critically examined.

MYTH 1: STUDENTS IN PACIFIC RIM COUNTRIES VALUE MATHEMATICS MORE THAN STUDENTS IN THE U.S.

Because students within the United States are not performing well on national assessments there is a tendency to believe that U.S. students do not value mathematics. On the other hand, since students in the Pacific Rim countries excel on the mathematics portions of these assessments, they must value mathematics. Although this may seem like a logical argument, the veracity of this myth should be challenged.

A study of TIMSS 2003 and TIMSS 1999 Video Study by Frederick K.S. Leung of The University of Hong Kong and Kyungmee Park of Hongik University in Korea (Leung, 2001), revealed that "students from the East Asian countries, other than Singapore, were all below the international norm in terms of their valuing of mathematics" (p. 4) and "that the percentage of East Asian students who indicated that they enjoyed learning mathematics was much lower than the international average" (p. 5). The study purports that, in general, "East Asian students held low value toward mathematics, did not enjoy learning mathematics, and had low self-confidence in learning mathematics" (p. 1). TIMSS-R results were consistent with former TIMSS findings. "With the exception of Singapore, all the top performing countries [Singapore, Korea, Chinese Taipei, Hong Kong, and Japan] had relatively negative attitudes toward mathematics.… [S]tudents from all the five top-performing East Asian countries had very low self-image of mathematics" (Kaiser, Leung, Romberg, & Yaschenko, 2002, p. 636).

A TIMSS study index, designed to measure the extent to which students value mathematics, ranked Japanese students at the bottom followed by Korea and Chinese Taipei. During a question and answer session at this conference, Ryosuke Nagaoka substantiated these findings reporting that Japan is currently faced with the challenge of reversing a trend of the diminishing value placed on the study of mathematics by students, their parents, and the entire Japanese society. He indicated that students in Japan are actually studying less and less, despite the efforts of teachers. Shigeo Yoshikawa (at this conference) expressed additional concerns regarding the trend. "We expect parents and people in society to get more interested in mathematic[s] education and to root for students learning mathematics … we expect teachers to get more interested in mathematics education." This heightened concern has impelled the Japanese government to focus on initiatives intended to help parents, teachers, and students understand the value of mathematics in their lives, hoping to reverse the current trend.

According to Sun Xiaotian, China is also experiencing resistance from students as they learn mathematics. Working groups formed from the Mathematics Curriculum Standards (MCS) initiative have suggested that "all students should learn mathematics, and school mathematics should be essential, valuable, and appropriate to students' need of learning" (Sun, p. 76, this volume). The MCS working group was comprised of mathematics professors, mathematics educators, mathematics teaching field practice coordinators, and classroom teachers. Their focus has been to create teaching, evaluation, and curricular material recommendations that can support the development of students' understanding of mathematics in a more personal way. For example, the teaching of proof is being reexamined. To accommodate the needs of students, Chinese teachers are presenting geometry, including proof, in a less formal way than in the past. According to Li Jun, learning mathematics in China will become more "attractive" by connecting mathematics to "students' living reality [using] contexts [and] challenges" (p. 127, this volume).

Hee-chan Lew shared that Korea is equally concerned with results from TIMSS and PISA that point to a problem with the affective aspect of mathematics education in Korea. Assessment data revealed that "affective characteristics were not friendly to mathematics compared with other countries" (Lew, p. 42 this volume). In fact, Korea's eighth graders were the lowest in confidence and interest towards mathematics. This has prompted many in Korea to question how mathematics can be made more valuable to students.

Although the TIMSS' value index ranked Singaporean students well above other East Asian students, according to Soh Cheow Kian, students in Singapore struggle with mathematics instruction and curriculum that is void of application. Instead, Singapore's students are interested in and would value studying mathematics that has application to their own lives. One of the current challenges facing the Ministry of Education in Singapore is to "make mathematics relevant and meaningful to all" and to develop curriculum that includes "meaningful applications and contexts."

Each of these Pacific Rim countries is concerned about the extent to which their students value mathematics and is working to understand the forces behind current trends. The United States is also interested in promoting the value of mathematics. The National Council of Teachers of Mathematics (NCTM) *Principles and Standards for School Mathematics* (2000) calls for a new vision of mathematics teaching including classrooms in which students "value mathematics and engage actively in learning it" (p. 3). Currently, students in the United States are scoring high in the affective domains measured on international assessments. The TIMMS' value index ranks U.S. students above students in all the Pacific Rim countries except Singapore. TIMSS and PISA survey questions, written to uncover attitudes about the study of mathematics, reveal that students in the United States seem to have an interest in mathematics and value what they are learning in their mathematics classrooms because it has application and relevance to their lives.

Ideally, students in all countries would find value in the study of mathematics. Curricular change initiatives being undertaken by the Pacific Rim countries may lead to improved results on affective measures. However, current international test indicators and reports by the Asian conference speakers support our claim of the mythical nature of the belief that Asian children value mathematics, while children in the United States do not.

MYTH 2: TEACHERS IN PACIFIC RIM COUNTRIES USE "TRADITIONAL" TEACHING STRATEGIES WHILE U.S. TEACHERS USE "REFORM" STRATEGIES

The use of "traditional" and "reform" in the statement of this myth is problematic because these words have different meaning to different people. Therefore, included here is a clarification of these terms for the purpose of this paper. Leung (2001) offers examples of "traditional" strategies in reference to mathematics teaching and learning in East Asian countries:

> Teaching is very traditional and old fashioned. Teachers in these countries seem to be ignorant about the latest methods of teaching, and think that competence in mathematics alone is sufficient for an effective teaching of the subject. Classroom teaching is conducted in a whole class setting, and given the large class size involved, there are virtually no group work or activities. Instruction is teacher dominated, and student involvement is minimal. Memorization of mathematical facts is stressed and students learn mainly by rote. There is ample amount of practice of mathematical skills, mostly without thorough understanding. (pp. 35-36)

In contrast, "reform" strategies, which according to this myth are more likely to occur in the United States, emphasize student-centered activities. The description of a mathematics classroom given by the NCTM (2000) will be used to characterize "reform" strategies:

> Students confidently engage in complex mathematical tasks chosen carefully by teachers. They draw on knowledge from a wide variety of mathematical topics, sometimes approaching the same problem from different mathematical perspectives or representing the mathematics in different ways until they find methods that enable them to make progress. Teachers help students make, refine, and explore conjectures on the basis of evidence and use a variety of reasoning and proof techniques to confirm or disprove those conjectures. Students are flexible and resourceful problem solvers. Alone or in groups and with access to technology, they work productively and reflectively, with the skilled guidance of their teachers. Orally and in writing, students communicate their ideas and results effectively. They value mathematics and engage actively in learning it. (p. 5)

It is not difficult to find additional support for Leung's conjectures regarding the "traditional" nature of teaching in East Asia. Rohlen, a researcher at Stanford (as cited in Fiske, 1983, July 13), found that "Japanese schools emphasize rote learning" (p. A1). Zhao (2005), in an article outlining the best and worst of the East and West regarding mathematics and science achievement, indicated that "they [East Asian countries] tried to teach too much and were knowledge- and teacher-centered" (p. 220). Similarly, Stigler and Perry (as cited in Wang & Lin, 2005) conducted a study of mathematics lessons in Japan, Taiwan, and the U.S. and found that "Chinese students spend substantially more time on learning activities led by their teachers than did their U.S. peers" (p. 6). On the Mathematically Correct Web site, similar statements are made: "The Japanese classroom does NOT emphasize group activity. The kids spend perhaps 20% of their time in groups ... the Japanese system is NOT "student-directed" or "student-centered" (Don't believe it!, 1997).

Leung (2001) not only described "traditional" teaching strategies in East Asia as outlined above, but also included descriptions of strategies common in the United States: "individualized teaching and learning is considered as the ideal ... consistent with well known underlying values such as child-centered education, constructivism, etc." (p. 44, 46). However, much research exists that provides evidence that, in spite of reform efforts, children in U.S. classrooms are not actually receiving instruction that could be characterized by the "reform" strategies described above. For example, Whittington (2002, December) in a report from a survey administered by Horizon Research, Inc. with responses from nearly 6,000 secondary mathematics and science teachers found

> a prevalence of traditional teaching practices (e.g., lecture, worksheets, reviewing homework). In each type of course, these practices were more common than any others.... The combination of whole class lecture/discussion and individual student activities (such as completing textbook problems) accounted for 62% of the time in a typical high school mathematics lesson; an additional 12% of instructional time is spent on noninstructional activities. (pp. 19-20)

Researchers in a separate study examining the TIMSS 1995 and 1999 Video Studies (Jacobs et al., 2006), concluded that "many of the features of teaching that show some alignment with *Principles and Standards*, whether in 1995 or in 1999, are features being implemented at the margins of teaching rather than at its core" (p. 30).

Teacher-centered instruction, as reportedly used in East Asian countries, is associated with "traditional" teaching practice; however, a growing body of research has indicated that this association may in fact be overly simplistic. For example, Grow-Maienza, Hahn, and Joo (2001) found that in spite of the fact that most classroom strategies were teacher-centered, "many more of the Asian teachers than teachers in the United States asked higher level questions, that is, involving computation in context, questions involving problem solving, and conceptual problems" (p. 364).

Additionally, information about classroom instruction shared and discussed by the Asian country representatives at the international conference exposed this as a second myth in need of reexamination.

China and Japan have similar classroom instructional formats. Both countries are currently involved in change initiatives that examine new ways of teaching that can support the development of creativity and innovation in students. Similar to the NCTM focus on process standards in *Principles and Standards for School Mathematics* (2000), China is focused on developing student thought processes with the hope that students will be more creative. China's national curriculum framework promotes teaching mathematics content "associated with social life ... and students' interests" (Sun) and recommends that "students should be encouraged to take active participation in investigation, field work, communication, and cooperation." In Japan, a *Hastimoto* is used to begin each lesson with a key question that provokes student's thinking on the mathematics to be examined that day, an instructional feature very similar to a "launch" activity included in some *Standards*-based curricula in the United States. As a whole group, with the teacher leading the discussion, students are given an opportunity to start thinking about the mathematics in the lesson and are encouraged to think creatively.

Singapore is engaged in initiatives to move away from their former focus on mathematics education as preparation for an exam to a new focus on mathematics education as preparation for life, a move toward "assessment for learning not learning for assessment" (Soh, p. 36, this volume). Mathematical problem solving is at the center of Singapore's curriculum framework with an emphasis on the development of processes like reasoning, communication, and making connections as intended learning outcomes. These changes are providing teachers in Singapore a new found autonomy, including making decisions regarding how they will teach students.

According to Lew, prior to the release of the new national curriculum, Korean "many math teachers do not consider mathematical application in daily life and reasoning ability to support logical conclusions as important goals for students to achieve" (p. 43, this volume). He shared the TIMSS results reporting that "89% of [Korean] students, which was the largest rate in all TIMSS countries, reported that mathematics classes were run by working together as a whole with the teacher teaching in the front." In contrast, "only 12% of [Korean] students, one of the lowest rates, worked in pairs or small groups with assistance from teachers" (Lew, p. 42, this volume). Korea's new curriculum, developed in part as a response to concerns raised by the TIMSS results, promotes instruction that allows students to "experience the joy of discovery and maintain their interest in mathematics" (Lew, p. 46, this volume). In the new curriculum, teachers are expected to build "mathematical power" by "focus[ing] on students' understanding of a problem and the problem solving process ... [and] focus on students' ability to think and solve problems in a flexible, diverse, and creative fashion" (p. 46, this volume).

The descriptions of classroom instruction shared by China, Japan, Singapore, and Korea reflect a philosophy more aligned with, than different from the United States. All five countries appear to be struggling with locating the balance between both teacher- and student-centered instruction, as well as, procedural and conceptual understandings of mathematics.

MYTH 3: STUDENTS IN PACIFIC RIM COUNTRIES RARELY USE CALCULATORS WHEN LEARNING MATHEMATICS WHILE AMERICAN STUDENTS OVERUSE CALCULATORS FOR COMPUTATION

The issue of when calculators should be introduced and how much students should use calculators has been a recurring battle in and out of the field of mathematics education since the invention of the calculator. Increasing the tension is the fact that research has been inconclusive as to whether calculators are helpful or harmful for students in learning mathematics (Hembree & Dessart, 1992; Loveless, 2004; Wilson & Naiman, 2004). However, the TIMSS and PISA studies found "that teachers in most countries, including the highest scorers, choose to limit calculator use in the primary grades more than U.S. teachers do" (Ginsburg, Cooke, Leinwand, Noell, & Pollock, 2005, pp. 23-24). Results from international comparisons have been used, by people with differing views on how to teach and learn mathematics, to suggest that students in the United States are overusing calculators, especially at the lower elementary grades. These same people propose that calculator use be limited as students learn to compute with numbers.

> Regarding basic number facts, Japanese students are much more competent than U.S. students. The facts are overlearned to a point where they are simply not a problem. Calculators are not allowed to replace oral or pencil and paper computations in lower elementary school where this competence is consciously sought and achieved nor are graphing calculators in widespread use at the secondary level. (Bishop, 1997, p. 242)

> Singapore specifically limits the use of calculators. Memorization of basic math facts is emphasized as a way to make mental calculations easier and

faster. While American schools give calculators to slower students, Singapore admonishes poorer-performing students to drill and practice more. (Izumi, 2000, p. 241)

These statements are supported by the results found in TIMSS and PISA, which report "only 31% of U.S. grade 4 teachers prohibit the use of calculators, but 53% of the teachers in the comparison countries disallow calculators" (Ginsburg et al., 2005, p. 24). There is variation among countries as to the degree that calculators are being used in classrooms.

Reports from the media use data from different international studies to exploit the calculator use in the classroom debate. For example, Richard J. Klutch stated, "the calculator was meant to make computation more convenient for people who already knew about numbers. Now, it threatens to crash the intellectual order, assuming the role of an end, when it is only a means" (1991, p. E17). The view of how and what calculators should be used for in the mathematics classroom is different depending on an individual's perspective of what mathematics is and how mathematics should be taught and learned. Klutch continues by stating, "education should be directed at children's heads, not their fingers, and while there are plainly legitimate use for calculators among advanced students, it is difficult to see that their widespread use in the elementary grades is wise" (p. E17). Recently, the Fordham Foundation released a report, *The State of the State of Math Standards* (Klein et al., 2005), which also claims that there is an overemphasis of calculator use within the United States as evidenced by state standards documents. However, a recent analysis of the emphasis and role of calculators in state standards documents found very little evidence to support the claim of such an emphasis, particularly in grades K–8 (Chval, Reys, & Teuscher, 2006). In fact, results indicate that the most common uses of calculators within state standards documents are for solving application problems/equations and for representing data—not for computation. This contradiction as to how and why calculators are being used has long been debated in the United States. In some classrooms, calculators are available and used extensively for many different learning activities: problem solving, graphing, and representing data. However, other classrooms neither have calculators nor have the resources to provide them.

In a conference session discussion of technology in the mathematics classroom, Li spoke to China's desire to follow in the footsteps of the United States. China wants the United States to be the forerunner in exploring and deciding the role and emphasis of technology. Once these decisions are made, China will follow closely behind. Japanese representatives at the conference espoused the view that all teachers should use technology widely and effectively. They believe that mathematicians need to be more broadminded about the use of technology and how it can be used in the classroom. Yoshikawa (p. 18, this volume) remarked that Japanese teachers "make use of *soroban*, calculators and computers when solving problems" but that the "use of calculators and computers in math class is up to the school and its teachers" (p. 18, this volume).

"Korean math teachers were the most conservative in using technology, such as the use of computers and calculators.... In the case of the eight grade (fourth grade), 76% (86%) of teachers and 93% (93%) of students never used computers in their classes. And 93% (96%) of teachers and 96% (92%) of students never used computers in their math classes." (Lew, pp. 42–43, this volume) However, Korea is trying to change this perspective with the recent publication of their seventh revised national curriculum. Students in Korea are being asked to use scientific and graphing calculators in their work and to engage in curricula that incorporates Excel spreadsheets. A major problem facing Korean teachers is not debating whether to use technology, but rather understanding how to use technology effectively especially in light of the Korean Ministry of Education's belief that technology should be introduced into the educational system earlier rather than later.

Singapore currently introduces scientific calculators in seventh grade and graphing calculator in 11th grade. However, recent discussions around the use of calculators in Grades 5 and 6 have created a debate within the Ministry of Education. The debate is fueled by members' fears that students' learning of more advanced mathematics may be handicapped by early introduction of calculators and that students may fail to learn "basic" facts. This same debate has been heard in the United States and it appears as though it will continue to be a topic of debate for educational institutions worldwide.

A common belief expressed across all countries, including the United States, is that calculators can be valuable tools if used appropriately. This belief, coupled with the information shared by each Asian country's Ministry of Education, challenges the notion that Asian students rarely use calculators in the learning of mathematics. In addition, a seminal study recently completed in the United States (Chval et al., 2006) suggests that U.S. students are not overusing calculators for computation. However, more research needs to be done to understand what is happening within the classroom. As evidenced at the international conference, other countries are looking to the United States for guidance in how to integrate technology, in particular calculators, appropriately in the classroom.

CONCLUSION

This paper challenges and critically examines three myths that have grown, in part, out of the results of international comparisons of student mathematics achievement. The myths addressed here are (1) students in Pacific Rim countries value mathematics more than students in the United States; (2) teachers in Pacific Rim countries use traditional teaching strategies while U.S. teachers use reform strategies; and (3) students in Pacific Rim countries rarely use calculators for learning mathematics while American students overuse calculators for computation. Analyzing and unpacking these myths revealed some underlying contradictions within each of them. The 2005 CSMC International Mathematics Curriculum Conference created a foundation on which further critical analyses can be built.

It was apparent from the conference that all five countries (China, Japan, Korea, Singapore, and United States) share similar dilemmas in the important issues of students' attitudes toward the value of mathematics, classroom instruction, and the use of technology. Additionally, all countries admitted looking to others for guidance in making the important decisions at hand. International collaboration and research can guide the direction of future discussions to benefit all countries as they make difficult decisions regarding their education system. The opportunities created by the conference to catalyze conversations and share strategies provided an important step toward fostering international understanding and cooperation in mathematics education—a step in the right direction.

REFERENCES

Bishop, W. (1997). *Math lessons from Japan: The TIMSS and the truth.* Retrieved February 27, 2006, from http://mathematicallycorrect.com/wbishop.htm

Chval, K., Reys, B., & Teuscher, D. (2006). What is the focus and emphasis on calculators in state-level K-8 mathematics curriculum standards documents? *Journal of Mathematics Education Leadership, 9*(1), 3-13.

Don't believe it! (1997). Retrieved February 20, 2006, from http://www.mathematicallycorrect.com/believe.htm

Fiske, E. B. (1983, July 13). Japan's schools: Intent about the basics. *The New York Times,* pp. A1, 28.

Ginsburg, A., Cooke, G., Leinwand, S., Noell, J., & Pollock, E. (2005). *Reassessing U.S. International mathematics performance: New findings from the 2003 TIMSS and PISA.* Washington, DC: American Institutes for Research.

Grow-Maienza, J., Hahn, D. -D., & Joo, C. -A. (2001). Mathematics instruction in Korean primary schools: Structures, processes, and a linguistic analysis of questioning. *Journal of Educational Psychology, 93*(2), 363-376.

Harms, W. (2005). *International conference on mathematics curriculum at the University of Chicago* [Press release]. Retrieved February 27, 2006, from http://www-news.uchicago.edu/releases/05/051108.math.shtml

Hembree, R., & Dessart, D. J. (1992). Research on calculators in mathematics education. In J. T. Fey & C. R. Hirsch (Eds.), *Calculators in mathematics education: 1992 yearbook* (pp. 23-32). Reston, VA: National Council of Teachers of Mathematics.

Izumi, L. T. (2000). Calculating the cost of calculators. *Capital Ideas, 5*. Retrieved June 23, 2008, from http://special.pacificresearch.org/pub/cap/2000/00-12-21.html.

Jacobs, J. K., Hiebert, J., Givvin, K. B., Hollingsworth, H., Garnier, H., & Wearne, D. (2006). Does eighth-grade mathematics teaching in the U.S. align with the NCTM standards? *Journal for Research in Mathematics Education, 37*(1), 5-32.

Kaiser, G., Leung, F. K. S., Romberg, T. A., & Yaschenko, I. (2002, August). *International comparisons in mathematics education: An overview*. Paper presented at the International Congress of Mathematicians, Beijing, China.

Klein, D., Braams, B. J., Parker, T., Quirk, W., Schmidt, W., Wilson, W. S. (2005). *The state of state math standards*. Washington, DC: Thomas B. Fordham Foundation.

Klutch, R. J. (1991, September 29). The calculator crutch. *New York Times*, p. E17.

Leung, F. K. S. (2001). In search of an East Asian identity in mathematics education. *Educational Studies in Mathematics, 47*(1), 35-51.

Leung, F. K. S., & Park, K. (2005, March/April). *Implications of TIMSS for mathematics curriculum reform in East Asia*. Paper presented at the Global Conference on Education Research in Developing and Transition Countries, Prague, Czech Republic.

Loveless, T. (2004, April). *Computation skills, calculators, and achievement gaps: An analysis of NAEP items*. Paper presented at American Educational Research Association Conference, San Diego, CA.

Mullis, I. V. S., Martin, M. O., Gonzalez, E. J., & Chrostowski, S. J. (Eds.). (2004). *TIMSS 2003 technical report*. Chestnut Hill, MA: TIMSS & PIRLS International Study Center, Lynch School of Education, Boston College.

National Council of Teachers of Mathematics. (2000). *Principles and standards for school mathematics*. Reston, VA: Author.

No Child Left Behind Act of 2001, Pub. L. 107-110, 115 STAT. 1425-2094.

Singham, M. (2003). The achievement gap: Myths and reality. *Phi Delta Kappan, 84*(8), 586-591.

Wang, J., & Lin, E. (2005). Comparative studies on U.S. and Chinese mathematics learning and the implications for standards-based mathematics teaching reform. *Educational Researcher, 34*(5), 3–13.

Watanabe, R., & McGaw, B. (2004). *Learning for tomorrow's world: First results from PISA 2003*. Retrieved March 16, 2006, from http://www.pisa.oecd.org/dataoecd/49/60/35188570.pdf

Whittington, D. (2002). *2000 national survey of science and mathematics education: Status of high school mathematics teaching*. Chapel Hill, NC: Horizon Research.
Wilson, W. S., & Naiman, D. Q. (2004). K-12 calculator usage and college grades. *Educational Studies in Mathematics, 56*, 119–122.
Zhao, Y. (2005). Increasing math and science achievement: The best and worst of the East and West. *Phi Delta Kappan, 87*(3), 219–222.

CHAPTER 23

REFLECTIONS ON ASSESSMENT

Angela D. Sutter
University of Missouri

Dana C. Cox and Karen L. Fonkert
Western Michigan University

Binding the images of thousands of tiny photographs into one large image is a photomosaic, an art form patented by Robert Silvers. Each tiny photograph is selected by a computer program on the basis of its subject, color, composition, and ability to contribute to a larger and often related image. These photomosaics grace the pages and covers of magazines such as the *Harvard Law Bulletin*, which in 1999 featured a photomosaic of one female graduate composed from thousands of yearbook photos of other female graduates (Silvers, 1999).

When viewed at a distance, photomosaics are instantly recognizable portraits of famous people, reproductions of well-known works of art, and images of familiar objects such as the globe. However, upon closer inspection, the once-clear image becomes fuzzy and ill defined. Features that were distinct from a distance blend at an arm's length. Coming even closer, it is impossible to see the larger picture, and one can only focus on the smaller photos, whose subjects only now become visible.

Mathematics Curriculum in Pacific Rim Coutries—China, Japan, Korea, and Singapore:
Proceedings of a Conference, pp. 351–360

As the events of the 2005 Center for the Study of Mathematics Curriculum International Conference unfolded, the complexity of studying mathematics curriculum, instruction and learning in a global sense was revealed. Layered like the photomosaic, what we know and want to know about mathematics curriculum, learning, and instruction looks quite different when viewed globally, nationally, or at the level of the individual. Assessment, a topic of discussion during the conference, is a powerful tool that has the ability to mold and shape the way the public views mathematics education and curriculum. Depending on the image provided by highly visible assessments and publications, people both in and outside of the mathematics education community draw conclusions, extrapolate future results, and make conjectures about what would change those projected results.

In this paper, we reflect on the purpose and nature of assessment in the United States by looking at assessment from a global perspective. As we consider how other nations are assessing their students' educational progress and how they are not, we notice some unique features of the U.S. system. Second, we reflect on the social and curricular effects of the variety of assessment practices utilized nationally and internationally as they were reported during the conference.

APTITUDE VERSUS ACHIEVEMENT

Throughout the International Conference, we heard statements from national representatives about the strong value placed on education by the citizens of China, Korea, Japan and Singapore. Parents in Korea spend $37 billion on extracurricular scholastic activities for their children, or about 10% of the family income (Lew, 2005a). In aggregate, this is $7 billion more than the Korean government spends in a year to educate the nation's students. As Lew summarized, "Most Korean parents believe their children's education is valuable enough for them to endure real hardships" (Lew, p. 37, this volume). The concern for the educational achievement of children is shared by parents in China. Sun (2005) shared, anecdotally, that Chinese parents feel more comfort when their children are seated at a desk studying than when playing outdoors.

The value that parents place on the education of their children is translated into direct pressure for students to achieve. Approximately 84% of Korean students attend extracurricular lessons where they are tutored and trained to pass stringent examinations (Lew, p. 39, this volume). Korean students are not alone in their attendance of extra lessons. Zang (1999, as cited in Kwok, 2004) studied such programs in Korea, Japan, and Taiwan and found that the curriculum of such programs was intended to bridge the daily school-based textbook curricula, and the high-stakes examination questions that students would encounter. As such, the motivation to attend *juku* or extracurricular mass-tutoring sessions comes from a desire to achieve high scores on examinations, get into more exclusive university programs, and climb the social ladder. Furthermore, Zang found that as the level of family income increased so did the participation rates in Japanese *juku* programs. In areas of high participation, *juku* themselves became stratified, the most elite of which accepting exclusively the highest achieving students. Thus, the programs intended to increase student achievement have the potential to further stratify the student population.

The programs that place such high stakes on exit examinations are contrary to the philosophy expressed by Soh (p. 36, this volume), "assessment for learning, not learning for assessment." By putting such emphasis on preparing students for examinations, students are placed in a position to value high examination scores rather than the mathematics they are learning. This leads to the question: Is the value that Korean, Japanese, and Singapore societies place on education really a value on achievement and competitive performance? To continue Soh's philosophy, are we "preparing students for the tests of life, not a life of tests?"

Achievement and competitive performance are not merely a function of student aptitude. In addition to socioeconomic factors mentioned previously, gender may also be a limiting factor for students hoping to achieve. For example, the male-dominated Korean culture steers Korean girls away from mathematical career paths, reserving these places for male students who are already expected to have higher mathematical achievement. Additionally, Korean and Chinese students who live in rural areas do not perform as well in mathematics as those educated in the cities (Lew, 2004). This poor performance is believed to be the result of inadequately prepared teachers and parental expectations.

This is not unlike the situation in the United States where students in poorer communities do not have the same opportunity to learn as students in communities with greater resources (Sirotnick, 2004). According to Oakes, Blasi, and Rogers (2004), "Absent the right resources, conditions, and opportunities, students cannot be expected to learn the knowledge and skills for which they, their teachers, and their schools are being held accountable" (p. 84). The existing gaps in achievement between gender and racial groups have also been well documented (e.g., Schoenfeld, 2002; Sullivan et al., 2005), leaving us to draw the conclusion that education in the United States, much like education in Pacific Rim countries, does not provide the same opportunities for all learners.

Assessment, then, reflects back these missed opportunities. However, with such high value placed on scoring well on national and international assessments, those students who are afforded the greatest opportunities are also offered the greatest rewards, thus perpetuating the gaps inherent in such a system.

Although standardized tests do exist for U.S. students, statewide examinations meant to monitor educational progress are not standardized and vary greatly (Sullivan et al., 2005). In Asian countries, high-stakes exams have a great deal more consistency than those given in the United States. National examinations in China, Japan, Korea, and Singapore serve as placement tools and as strict sorting criteria for university acceptance since they provide a uniform comparison of student achievement. Student performance on exams determines which track the students will travel during preparatory school and which university they will attend upon graduation. The exams have high stakes, with performance determining the social and economic future of each student who takes them.

Variations are seen in United States high-stakes high school exit exams, which individually serve the same purpose as exit exams in the aforementioned nations, but present unique assessed curricula which provide challenges to curriculum designers, and those interested in making interstate comparisons of students (Sullivan et al., 2005). The lack of a national curriculum—a shared focus on what our students are to learn—has created a fragmented system of standards that guide multiple and varied student assessments in the United States. This is evidenced by the variations in state curriculum documents (Reys, Lappan, & Kim, 2005) and the various state assessments created to determine student learning of the curriculum (Sullivan et al.).

With the passage of No Child Left Behind (NCLB) legislation in 2001, states are required to increase the number of assessments given to students, including administration of an assessment at the high school level. While not all states will mandate passage of the exam as a means for high school graduation, many states are rethinking existing exams in light of NCLB legislation. Currently 19 states have high school exit exams in place. By 2012, current legislation indicates this number will increase to 26. As the number of states mandating exit exams grows, so does the population of students affected by the exams. In 2005, 50% of U.S. public school students attended school in a state with an exit exam. By 2012, that percentage will increase to 72%, with the largest percentage of these students being minority (82%), special education (71%), and English language learning (87%) students—those at greatest risk of passing the exams (Sullivan et al., 2005).

These exams were initially created to appease society's outcry that our public school students were ill-prepared for life after high school. With the push for higher standards for high school graduates, many of the states claim their tests are not minimum competency exams, but standards-based exams that hold students to the higher standards. In fact, 16 of the 19 states currently administering exams claim that their exams are standards-based, that is to say, hold higher expectations for students than previously used minimum-competency exams. However, a report by Achieve, Inc. (American Diploma Project, 2004) examining exit exams in six states found that the content being assessed was of low quality and was not aligned with the knowledge a student would need to enter the workforce or college upon high school graduation.

The inability to create a functioning assessment has caused the seriousness of these exams to be taken into consideration (Lewis & Lee-Bayha, 2003), leading to weak public support for high-stakes assessments. For example, California, which had an exit exam in place in 2003, has since withdrawn it with plans for redevelopment and reinstitution of the exam in 2008. Similarly, Arizona has, for the fourth time since 1996, withdrawn the requirement of student passage of an exit exam to receive a high school diploma (Sullivan et al., 2005). Without consistency in implementation of these exams, they will continue to lose public support (Lewis & Lee-Bayha, 2003), diminishing their importance as gauges of student knowledge.

ADDITIONAL ACCOUNTABILITY

In the United States as well as the Pacific Rim countries represented at the CSMC conference, large-scale testing is also used to assess the quality of mathematics curriculum. National tests such as the National Assessment of Educational Progress given in the United States and similar tests used in other countries (e.g., National Assessments given in Japan; Yoshikawa, 2005) provide aggregate results to assess the quality of mathematics education. Other statewide tests such as the Michigan Educational Assessment Program and the Texas Assessment of Academic Skills are used to assess the quality of state programs. These tests monitor student achievement, however results are presented in such a way that they extend the accountability beyond individual student performance and a statewide report. They are often used to hold numerous parties accountable: schools, teachers, and students. Annual Yearly Progress, a mandate of NCLB that is tied to student achievement testing, places pressure on school personnel whose job security is tied to school performance (Cox, at this conference). However, states hold different requirements for students. Which state one lives in is a powerful determiner of what is deemed as satisfactory achievement, as different states have different assessments (Schmidt, at this conference) tied to different curricula.

The disaggregating of these types of results by classroom, school, and district is unique to the United States. These kinds of results are not provided by the Japanese ministry of education (Yoshikawa, at this conference). Their intent is not to force schools and teachers to compete against one another, but to influence them to conform as closely as possible to national guidelines for mathematics instruction. That the United States highlights the performance not only of individuals, but of schools, districts and states in a way that encourages comparison of results yields high anxiety and pressure not only for students, but for teachers, schools, and states. Cox (at this conference) discussed clearly that these pressures impact mathematics curriculum on many levels. Grade level expectations are set by state boards who determine the ideal mathematics curriculum. Curriculum designers are beholden to many such state expectations when designing materials to facilitate learning. Teachers are forced to make curricular decisions as preparing students for the examinations falls within their duties. Lastly, public perceptions of mathematics education have strong influences on the curricular decisions made by school districts.

ASSESSMENT AND EDUCATIONAL ADVANCEMENT

In the United States, the use of technology during assessment is promoted. In its *Principles and Standards for School Mathematics*, the National Council of Teachers of Mathematics, or NCTM, (2000) asserts that "technology aids in assessment, allowing teachers to examine the processes used by students in their mathematical investigations as well as the results, thus enriching the information available for teachers to use in making instructional decisions" (p. 26). NCTM also states in *Mathematics Assessment: Myths, Models, Good Questions, and Practical Suggestions* (Stenmark 1991), "Students must be permitted to use calculators [on standardized tests]; as the use of these 'fast pencils' become routine in classrooms, evaluation must keep pace" (p. 9). In *Assessment Standards for School Mathematics* NCTM (1995) gives an example of a classroom, in which a statistical package and graphing software are used by students to investigate statistical concepts, and to give a presentation of their findings to the class. Graphing calculators, dynamic geometry software, statistical software, spreadsheets, and computer algebra systems are all examples of types of technology that are desirable for use in assessment in the United States.

As Heid (at this conference) and Chval (at this conference) stated on the last day of the conference, there is still progress to be made in the area of technology use in the United States. However, there is some evidence of technology being utilized in conjunction with assessment. Certain standardized tests such as the ACT and SAT allow students to use graphing calculators. Some colleges allow students to use calculators on their placement tests. Furthermore, since the advent of reform mathematics curricula, many classroom teachers allow students to use various forms of technology on assessments. In American classrooms, students are doing performance assessments, projects, and presentations using technology as well as utilizing graphing calculators on classroom assessments (Bethell & Miller, 1998; Coxford & Hirsch, 1996; Dugger, 1999).

Reform math curricula have played a major part in increasing the use of technology on assessments in the United States. Instruction and assessment need to be aligned and integrated, promoting the alignment and integration of technology and instruction (NCTM, 1995). "For years we have understood that the best test also instructs and the best instructional tasks are rich diagnostic opportunities. It is time to use these understandings in all aspects of instruction and assessment" (Stenmark, 1991, p. 5). Therefore, when students using reform curricula, are learning mathematics through rich tasks that utilize technology, a teacher can better assess what the students know (NCTM, 2000). Furthermore, since those students are using technology during their instructional time, the technology is also used in their assessments so that the assessment is aligned with the instruction.

Comments made during the conference and publications written by international representatives suggest that the other countries represented at the conference are not as far along in incorporating technology into assessment. In Singapore, they "still do not allow the use of calculators in primary schools" and "do not yet use graphic calculators in junior colleges" (Lee, 1999, p. 190). Lee continued, sounding hopeful that Singapore will continue to incorporate more technology, "We want to train a work force which can meet the new challenge in an information age" (p. 192). Singapore has a master plan to get computers into their schools, and technology into the curriculum, "want[ing] to develop a culture in which the use of instructional technology is commonplace" (p. 193). At the conference, Lee (at this conference) talked about the progress they have made on this master plan. Now that the technology is working its way into Singapore schools, the current focus is on the changes in curriculum and assessment needed to incorporate the technology. Their goals are to use technology in the classroom, and on examinations.

A similar position is taken in China, where most students have never used a graphing calculator or a spreadsheet, but they have used a scientific calculator and geometry software. Li Jun (at this conference) said at the conference that they currently have a national project underway to find a way for teachers and students to use calculators and computers in instruction. She also said that the junior high uses technology, but that it is not allowed in the examination. Shanghai is discussing using calculators in the examination, and calculators are allowed on university entrance exams.

According to Nagaoka (at this conference), Japan is very open to technology use in instruction. At all levels, their curriculum promotes the use of a calculator, and it is not difficult for them to get graphing calculators into the classrooms. Also, he said that the government is eager to introduce computers into the classroom. However, he seems to think that the electronic dictionaries that are now quite prevalent, and other electronic devices of the future, might be more beneficial. In some junior high schools, using calculators is allowed even in exams. This practice is even encouraged in some schools, according to Nagaoka.

Lew of Korea said at this conference that all of their textbooks must contain some problems that use technology, but this may not reflect teacher values. Reluctant to embrace technology, teachers tend to ignore technology recommendations. He also said that the national curriculum encourages computer use, but all the content can be learned without a computer. Because it is not required, and perhaps because teachers are ignorant of how to best incorporate technology into lessons and assessment, there has not been great progress in this area. Thus, while he has a positive attitude toward technology use in instruction and assessment, his comments suggested that most Korean teachers do not share his view.

The representatives from China, Japan, Korea, and Singapore spoke about goals for using technology similar to the United States' goals. Japan's Nagaoka was actually surprised to hear that not all teachers in the United States are making full use of technology, since he had heard the opposite. China's Jun said that they follow the United States on this topic. Thus, the other countries are paying attention to the way the United States intends to incorporate technology into instruction and assessment. Therefore, we should carefully consider the example we set.

CONCLUSION

The implications of assessment on the development and study of mathematics curriculum are numerous. As assessments are completed and added to the register of academic results, they cannot help but shape the public's awareness and perception of mathematics education. Assessment is a powerful tool that can be used to manipulate and influence the field. As many of the speakers noted, in order to change the content and the method by which mathematics is taught, the most expedient vehicle is to change the assessment. "No reasonable person is against accountability that enhances the quality of education. But reasonable people can and do differ regarding what an accountability system needs to entail if it is to help us achieve the shared goal of improving education" (Linn, 2003).

To simply add more assessment to current programs of education would do nothing but exacerbate the current anxiety felt around the world as systems are altered based on fragmented and contradictory results. As a community, the field of mathematics educators should examine the current state of assessment and reevaluate the methods by which stock is taken of student progress. Particularly in the United States, some consensus could be reached regarding the purpose and methods of examining students. This consensus would bring consistency to student examinations, thereby increasing the interstate consistency of the mathematics that is taught and learned.

This is not to imply that we need to settle on one national method of teaching and assessing students. Like the photomosaic, learning mathematics is complex, and cannot be fully studied from one vantage point. What we can learn from large-scale international study of students and teaching is of no less value than what we can learn from daily, embedded assessments given by teachers in individual classrooms. What is important is putting together the infrastructure by which all forms of assessment can work together, providing a complete picture of student progress and of our educational system. How can assessments exist in a coherent and multidimensional context rather than as mutually exclusive contradictions that do nothing but cloud the public perception of mathematics education? How can we create a photomosaic of student learning?

REFERENCES

American Diploma Project. (2004). *Do graduation tests measure up? A closer look at high school exit exams*. Washington, DC: Achieve.

Bethell, S. C., & Miller, N. B. (1998). From an E to an A in first-year algebra with the help of a graphing calculator. *Mathematics Teacher, 91*(2), 118–119.

Coxford, A. F., and Hirsch, C. R. (1996). A common core of math for all. *Educational Leadership, 53*(8), 22–25.

Dugger, W. E., Jr. (1999). Technology for all Americans: Providing a vision for technological literacy. *The 3 R's for Teachers: Research, Reports, & Reviews—A Publication of The MASTER Teacher, 9*(3), 1–4.

Kwok, P. (2004). Examination-oriented knowledge and value transformation in East Asian cram schools. *Asia Pacific Education Review, 5*(1), 64–75.

Lee, P. Y. (1999). Mathematics education in Singapore. In Z. Usiskin (Ed.), *Developments in school mathematics education around the world: Proceedings of the fourth UCSMP international conference on mathematics education* (pp. 188–193). Reston, VA: National Council of Teachers of Mathematics.

Lew, H. C. (2004, July). *Mathematics education in Korea after TIMMS*. Paper presented at the Tenth International Congress on Mathematical Education, Copenhagen. Retrieved June 23, 2008, from http://www.mathlove.com/new3/notice/data/Chap01(Lew).pdf

Lewis, J. W., & Lee-Bayha, J. (2003). *Making sure exit exams get a passing grade*. San Francisco: WestEd. Retrieved June 23, 2008, from http://www.wested.org/cs/we/view/rs/700

Linn, R. L. 2003. *Accountability: Responsibility and reasonable expectations*. Los Angeles: Center for the Study of Evaluation, National Center for Research on Evaluation, Standards, and Student Teaching.

National Council of Teachers of Mathematics. (1995). *Assessments standards for school mathematics*. Reston, VA: Author.

National Council of Teachers of Mathematics. (2000). *Principles and standards for school mathematics*. Reston, VA: Author.

Oakes, J., Blasi, G., & Rogers, J. (2004). Accountability for adequate and equitable opportunities to learn. In K. A. Sirotnik (Ed.), *Holding accountability accountable: What ought to matter in public education* (pp. 82–199). New York: Teachers College Press.

Reys, B., Lappan, G., & Kim, O. K. (2005, April). *State curriculum standards: Do they represent a national curriculum?* Paper presented at the National Council of Teachers of Mathematics annual meeting and exposition, Anaheim, CA.

Schoenfeld, A. H. (2002). Making mathematics work for all children: Issues of standards, testing, and equity. *Educational Researcher, 31*(1), 13–25.

Silvers, R. (Artist). (1999). *Sheila Flynn '01 and 1000 women from the classes of 1953–98*, photomosaic based on photograph by Farnsworth Blalock Photography. Retrieved June 23, 2008, from http://www.law.harvard.edu/alumni/bulletin/backissues/spring99/cover.html

Sirotnik, K. A. (2004). Holding accountability accountable—Hope for the future? In K. A. Sirotnik (Ed.), *Holding accountability accountable: What ought to matter in public education* (pp. 148–170). New York: Teachers College Press.

Stenmark, J. K. (Ed.). (1991). *Mathematics assessment: Myths, models, good questions, and practical suggestions*. Reston, VA: National Council of Teachers of Mathematics.

Sullivan, P., Yeager, M., Chudowsky, N., Kober, N., O'Brien, E., & Gayler, K. (2005). *States try harder but gaps persist—High school exit exams 2005*. Washington, DC: Center on Education Policy.

AFTERWORD

Zalman Usiskin

Every mathematics teacher knows that what students learn in mathematics is highly dependent on the taught curriculum. Researchers call this variable "opportunity to learn" and have further refined it with such phrases as "time on task" and "active engagement." Since the taught curriculum is known to be highly dependent on the textbook curriculum, and the textbook curriculum is acknowledged to be dependent on national reports and often on state guidelines, what has been termed the ideal curriculum, it naturally follows that all of these curricula are important. They determine what students learn. Those of us who study mathematics curriculum believe strongly in the centrality of curriculum in any examination of mathematics learning and teaching.

I spent most of my academic career in a department of education known for its research and consideration of education as a social science. Some years ago, the logic of the preceding paragraph was challenged by a departmental colleague who put forth the notion that the textbook decisions of school districts are very much influenced by the socioeconomics of the district. While today I do not remember the particular study or even the name of the colleague, I do remember being shaken by the notion that the textbook, an independent variable to me, was a dependent variable to him.

Mathematics Curriculum in Pacific Rim Coutries—China, Japan, Korea, and Singapore: Proceedings of a Conference, pp. 361–362

The papers found in this volume reaffirm both these views of curriculum. How students perform in mathematics in China, Japan, Korea, and Singapore is, without doubt, a function of the textbooks from which they are taught and the textbooks are required to follow the national curriculum guidelines. But economic factors also play a major role in what students learn. The out-of-school industry in Korea compels us to consider how much time students spend learning mathematics outside of class when we look at possible causes of differences in performance among students, schools, or countries. The very high socioeconomics of Singapore compared to its neighbors, fueled by a labor force which in 2002 was 40% unskilled with less than secondary education, made up largely of foreign nationals that constituted 29% of the total work force (Yap, 2003) and yet whose children are not tested as Singaporeans, compels comparisons not with our entire nation, but with our affluent suburbs with correspondingly high socioeconomics compared to their neighboring cities.

The papers of the U.S. researchers in this volume demonstrate that the study of comparative curriculum is a very interesting and fruitful area, worthy of continued examination. The analysis of mathematics curricula from around the world is still in its nascent stages, having begun less than a half-century ago, and we seldom see analyses using the same kinds of tools. This is an area ripe for canonical forms of research.

The panels at this conference were designed to force the discussion into practical issues of particular prominence in discussions of mathematics education today. What are, could, and should be the roles of technology in the learning and doing of mathematics? How can we aid performance in algebra and geometry? Even the brief remarks from the U.S. speakers on these questions indicate the complexity of these issues, and the reactions from the doctoral students reiterate that there are countless interesting and valuable problems to study in mathematics curriculum.

REFERENCE

Yap, M. T. (2003). Singapore. In *Migration and the labour market in Asia: Recent trends and policies* (pp. 368-371). Paris, OECD.

Printed in the United States
127456LV00002B/4/P